AF358691

Municipal Wastewater Management

Municipal Wastewater Management

Editor

Charikleia Prochaska

MDPI • Basel • Beijing • Wuhan • Barcelona • Belgrade • Manchester • Tokyo • Cluj • Tianjin

Editor
Charikleia Prochaska
Chemistry
Aristotle University of
Thessaloniki
Thessaloniki
Greece

Editorial Office
MDPI
St. Alban-Anlage 66
4052 Basel, Switzerland

This is a reprint of articles from the Special Issue published online in the open access journal *Sustainability* (ISSN 2071-1050) (available at: www.mdpi.com/journal/sustainability/special_issues/ Municipal_Wastewater_Management).

For citation purposes, cite each article independently as indicated on the article page online and as indicated below:

LastName, A.A.; LastName, B.B.; LastName, C.C. Article Title. *Journal Name* **Year**, *Volume Number*, Page Range.

ISBN 978-3-0365-1674-5 (Hbk)
ISBN 978-3-0365-1673-8 (PDF)

Contents

About the Editor

Charikleia Prochaska

Dr. Charikleia Prochaska holds a Chemistry degree, a Master of Science (MSc) in Chemical Technology, and a PhD in Environmental Technology from the Aristotle University of Thessaloniki, Greece. She has participated for over ten years in several research projects in the fields of Wastewater Treatment and Management, in the broader fields of Chemical and Environmental Technology and she is also experienced in surface characterization using Atomic Force Microscopy. She has extensive teaching experience in the fields of Chemistry and Environmental Science and Technology, at both under-graduate and post-graduate level. As of April 2014, Dr. Prochaska works as a Laboratory Teaching Staff in the Chemistry Department of the Aristotle University of Thessaloniki, Greece. Dr. Prochaska has authored or co-authored several papers and chapters in edited books, and has long been involved as a reviewer and, from October 2019, serves as a reviewer-board member of Sustainability Journal.

Preface to "Municipal Wastewater Management"

This Special Issue of *Sustainability* Journal belongs to the Section of Social Ecology and Sustainability. It was developed by the Journal and was invited to be led through the process of guest-editing, as part of the Journal's growing series of thematic issues.

Under the broad umbrella of its title "Municipal Wastewater Management" eleven peer-reviewed papers were published, covering a wide thematic range of both research and review articles.

The eleven papers form a collection that adds to our understanding of "Municipal Wastewater Management" in ways that would hopefully trigger readers' interest and stimulate colleagues towards advancing their research and overcoming weaknesses and limitations.

All authors are highly acknowledged for contributing to the realization of this Special Issue.

Charikleia Prochaska
Editor

 *sustainability*

Editorial

Special Issue: Municipal Wastewater Management

Charikleia Prochaska 

Laboratory of Chemical and Environmental Technology, School of Chemistry,
Aristotle University of Thessaloniki, 54124 Thessaloniki, Greece; prohaska@chem.auth.gr

check for
updates

Citation: Prochaska, C. Special Issue:
Municipal Wastewater Management.
Sustainability **2021**, *13*, 7588. https://
doi.org/10.3390/su13147588

Received: 1 July 2021
Accepted: 3 July 2021
Published: 7 July 2021

Publisher's Note: MDPI stays neutral
with regard to jurisdictional claims in
published maps and institutional affiliations.

Municipal wastewater management is a well-established field in most parts of the world nowadays. However, there are still many challenges that need to be faced, not only in developing countries, where municipal wastewater is often discharged without previous treatment, but also in developed countries, where wastewater treatment plants have been employed for decades, with the goal of reaching the stage of highly advanced sustainable municipal wastewater management that would protect both the environment and humans' health, be energy-, water-, and cost-efficient, and have the ability to adapt to the current environment and the needs of the community.

In that sense, special issues, focused on the particular subject of "Municipal Wastewater Management" are always relevant and worth being reintroduced, as they bring together scientific aspects of the particular subject that might have not been revealed or met, unless such a special issue had been developed. This particular issue had the chance to bring together, after a rigorous peer review process, five original research papers and four review articles that focus on municipal wastewater management at a broad range of scales, that is: from practical local actions to global principles.

The aim of this editorial is to point out key aspects of the guest-editing process and to highlight the contributions, in a way that: (a) their importance is enhanced, and (b) how they link together, under the umbrella of this special issue, is defined, so that all contributions can stimulate a broader scientific audience, awareness and understanding.

Starting by highlighting the original research papers, contributed to this Special Issue, Fytianos et al. (Contribution 1) in "Biocorrosion of Concrete Sewers in Greece: Current Practices and Challenges", address a topic that exists in the majority of aged municipal wastewater collection systems, where concrete was used before polyvinyl chloride (PVC) pipes became common practice in the mid-1980s. According to estimations, the biogenic corrosion of concrete sewer pipes, due to the production of H_2S in sewage, represents approximately 10% of the total sewage treatment cost. In their holistic approach, Fytianos et al. administered a questionnaire survey to stakeholders working in municipal wastewater treatment plants serving more than 50% of the total country's population, and validated the survey answers with field measurements and analyses. With their work, not only is the nature and extent of concrete biocorrosion problems in Greece presented for the first time, but also the need for holistic approaches to handle complex aspects of municipal wastewater management, such as biocorrosion in old sewer networks, is highlighted, providing a broader research interest in this work.

As the holistic approach to deal with biocorrosion in old municipal sewer networks cannot exclude the integration of both technical and financial information in applying sustainable maintenance methodologies to the sewage network, in their follow-up paper entitled "Least Cost Analysis for Biocorrosion Mitigation Strategies in Concrete Sewers", Fytianos et al. (Contribution 2) use the same research area as before and this time present results from a cost-comparative analysis, focusing on an annuities calculation for the evaluation of microbiologically induced corrosion or biocorrosion mitigation methodologies, used in the maintenance of concrete sewers. They conclude that spraying with magnesium hydroxide slurry is the most advantageous maintenance technique, based on direct and indirect economic assessment. As each case is different, a universal tool for the economic evaluation of various biocorrosion mitigation strategies could be useful. In that sense, the authors' future work, focusing on the development of such a universal financial assessment tool, is worth being followed.

The application of SCADA (Supervisory Control and Data Acquisition) systems to the remote monitoring and controlling of wastewater treatment plants is the focus of Brad et al.'s (Contribution 3) publication entitled "Lifecycle Design of Disruptive SCADA Systems for Waste-Water Treatment Installations". The major contribution of this paper is a proposed structured methodology for optimizing SCADA systems from a lifecycle perspective for their application to the specific case of wastewater treatment plants. This kind of approach was not previously reported in the literature for this stage of design. The theoretical model developed by the authors is tested under practical conditions, illustrating both its applicability and its limitations.

García-López et al.'s (Contribution 4) publication entitled "The financing of wastewater treatment and the balance of payments for water services: Evidence from municipalities in the Region of Valencia" casts light on the issue of how tax rates can provide the necessary revenue to finance the cost of urban wastewater treatment, using evidence from municipalities in the region of Valencia, in south eastern Spain. The analysis provided in this work demonstrates the importance of an appropriate wastewater treatment tariff structure that would include aspects, such as water pollution and energy costs, that are not currently intergraded into wastewater treatment tariffs, not only in Valencia but in most municipalities within Europe and around the world. Through the local insights, provided in this work, we can gain an understanding of how wastewater treatment taxes can lead to the economic sustainability of urban wastewater treatment plants, a matter of great importance and challenge for all urban wastewater services worldwide.

Tsiakiri et al.'s (Contribution 5) publication entitled "Estimation of Energy Recovery Potential from Primary Residues of a Municipal Wastewater Treatment Plant" aims towards the identification of the biomethane potential of non-conventional sources derived from municipal wastewater treatment processes. In this work, byproducts deduced from the primary treatment process stage were collected from four sewage treatment plants in Greece and analyzed for their solid and fat content, as well as their concentration of dissolved organic matter and nutrients, and were subjected to anaerobic digestion treatment for the measurement of their biomethane production potential. The highest potential for biogas utilization was found in screenings collected from a treatment plant receiving wastewater from an area with combined rural and agro-industrial activities. Floatings from grit chambers presented the smallest potential for energy recovery. However, these wastes were found to be potentially suitable for energy production, in secondary sludge co-digestion units. Wastewater treatment plants can benefit from such research approaches, aiming to minimize the energy consumption of the municipal wastewater treatment process, leading even to the production of power in wastewater facilities.

Psaltou et al. (Contribution 6) in "Effect of Thermal Treatment on the Physicochemical Properties of Minerals Applied to Heterogeneous Catalytic Ozonation" examine the effect of thermal treatment on three inexpensive minerals, i.e., zeolite, talc, and kaolin (clay), which present different physicochemical properties as potential catalysts for the removal of para-chlorobenzoic acid (p-CBA), a typical micro-pollutant, commonly used as a model compound to indirectly evaluate the production of hydroxyl radicals in ozonation systems. The addition of an appropriate catalyst can enhance the efficiency of the ozonation process, which is a promising treatment technique, especially for the removal of micro-pollutants found in municipal wastewater treatment plants. Thus, such research is always of interest, and the results presented here are promising, especially regarding the catalytic activity of talc (primarily) and kaolin (secondarily), which are not as widely researched as zeolite, as potential catalysts for enhancing the efficiency of heterogeneous catalytic ozonation, applied to municipal wastewater treatment processes.

Dang et al. (Contribution 7) in their work "Loofah Sponges as Bio-Carriers in a Pilot-Scale Integrated Fixed-Film Activated Sludge System for Municipal Wastewater Treatment" bring focus on another interesting subject of municipal wastewater treatment, especially for developing countries: the use of natural materials as carriers for the integrated fixed-film activated sludge system (IFAS). The authors apply modified loofah sponges, as bio-carriers, in a pilot-scale IFAS for the treatment of real municipal wastewater. Loofah is an annual herbaceous plant from the cucurbitaceous family, and its fully developed fruit is the source of the loofah scrubbing sponges normally used in bathrooms and kitchens. The results presented here are encouraging and worth being noticed, as modified loofah can serve as a possible replacement to the expensive and environmentally unfriendly polyethylene bio-carries for municipal wastewater treatment applications in developing countries, where loofah is locally abundant.

The concept of applying hybrid systems of constructed wetlands (CWs) for the treatment of the domestic wastewater of small communities has been studied for many years now. However, this alternative technology has not been widely accepted, mainly because its application has not been accompanied with a parallel effort to gain wider community acceptance of it over the conventional wastewater treatment systems. Lavrnić et al. (Contribution 8) in "Potential Role of Hybrid Constructed Wetlands Treating University Wastewater-Experience from Northern Italy" place their research in the University campus of Bologna (Northern Italy) and attempt to assess the potential of their hybrid CW system to be used in universities for wastewater treatment by incorporating the public's opinion (students and university staff) in their evaluation. The results, presented in this research, show that the hybrid system met the Italian limits for discharge in natural water bodies and some of the limits for wastewater reuse in Italy and the European Union. The positive attitude towards CWs and wastewater reuse, found among the survey participants, reinforce the belief that hybrid CWs (planted and unplanted) can be considered as a feasible technology for application at universities.

The Special Issue had also the chance to bring together one review and two mini-review articles, which focus on summarizing and presenting, in different perspectives, the current state of research concerning, respectively, the following topics: (a) the history, development and future challenges of urban wastewater treatment in Greece; (b) solar photocatalysis for emerging micro-pollutants abatement and water disinfection; (c) the policies regarding the ocean dumping of treated municipal sewage in the Republic of Korea.

The mini-review of Prochaska and Zouboulis (Contribution 9), entitled "A Mini-Review of Urban Wastewater Treatment in Greece: History, Development and Future Challenges" serves as another example of how the municipal wastewater management of a particular country can be of interest to a wider readership, as it addresses the country's own wastewater management path in respect to the global thinking around sustainable environmental development. In this review paper, the authors revisit the development history of Greece's municipal wastewater management, highlight the future needs of sustainable development, explore Greece's own wastewater management path, and look towards the future from several aspects, including the European Union's policies and international technological trends, in order to identify wastewater treatment plants not only as sites of pollutant removal, but also as places where energy is efficiently used and environmental sustainability is being practiced. The review refers also to the promising area of ongoing research on COVID-19 which involves using sewage to monitor virus circulation in communities and to detect possible outbreaks, even before clinical cases have been identified. As novel enveloped viruses are expected to emerge in the future, integrating this approach with proper onsite wastewater management could help tackle the problem, at the local scale, and avoid larger-scale virus outbreaks.

As stated by Venieri et al. (Contribution 10), the occurrence of emerging micro-pollutants in the aquatic environment, as well as the presence of various pathogenic microorganisms, impose the application of effective purification methods in order to maintain high hygiene standards and protect public health. Venieri et al. in "Solar photocatalysis for emerging micro-pollutants abatement and water disinfection: A mini-review" give a

Sustainability **2021**, *13*, 7588

clear overview of the recent progress in the field and point out important considerations in the design of such systems. Lab and pilot-scale applications are presented, current trends regarding the elimination of antibiotic resistant bacteria, and resistance genes by means of solar photocatalysis are discussed, with a view to investigating the prospect of using those purification methods for the control resistant microbial populations found in the environment. Understanding the interactions of the various water components (both inherent and target species) is key to the successful operation of a treatment process and its scaling up.

Treated sewage sludge disposal is always a matter of research interest, as policies and regulations worldwide are aiming towards a sustainable sludge disposal strategy, with minimum sludge disposal to landfill, maximum energy recovery, agricultural uses and no ocean dumping. In this context, the review article of Chung et al. (Contribution 11) entitled "Overview of the Policies for Phasing out Ocean Dumping of Sewage Sludge in the Republic of Korea" brings focus to the approach followed in the Korea Republic regarding the evolution of policies for the disposal of treated municipal wastewater sludge into the ocean, which was stopped in the Republic of Korea in 2012. The article helps in understanding how international conventions, such as the Convention on the Prevention of Marine Pollution by Dumping of Wastes and Other Matter, commonly called the "London Convention", entered into force in 1972 and replaced in 1996 by the "London Protocol", can gradually impact the swiftness of local policies towards environmental protection.

Taken the contents of this Special Issue as a whole, it is clear that "Municipal Wastewater Management" is an ongoing field of research with the ability to incorporate current environmental and humans' health challenges into its analyses. The use of municipal sewage to monitor COVID-19 virus circulation in communities and the estimation of possible outbreaks, even before clinical cases have been identified, is another fact that justifies this.

In light of the coronavirus pandemic, interest in the impact that research on municipal wastewater management can have on improving humans' health and protecting the environment is being rethought. In respect to this, there is an essential need for scientific publications that present varieties of case studies and discuss best practices, in order for wastewater treatment plants to be seen not only as sites of pollutant removal but also as places where energy is efficiently used and environmental sustainability is being practiced, in close relation to the needs of the community.

Viewed in this way, the papers collected in this Special Issue aim to reach a broad readership that can gain awareness and understanding of their topics and be stimulated into future research and collaborations that would improve all stakeholders' engagement towards promoting sustainable municipal wastewater management.

Funding: No external funding was received.

Acknowledgments: This Special Issue was developed following on from the invitation sent by Elaine Li, Managing Editor at MDPI. The Editor-in-Chief of Sustainability Journal and Section Editors of Social Ecology and Sustainability are acknowledged for providing me with the opportunity to lead this Special Issue as guest-editor. Acknowledgements are owned to the Editorial Board members, who served as Academic Editors, when needed, and to the anonymous reviewers, who both devoted much of their precious time in reviewing the submitted manuscripts and in helping the best of them being published. All authors are highly acknowledged for offering their valuable contributions. The Managing Editor, the Assistant Editors and all the members of the Editorial Office, are sincerely acknowledged for providing invaluable support throughout the editing process. Last but not least, the publisher MDPI, is highly acknowledged for allocating resources towards the realization of this Special Issue.

List of Contributions:

1. Fytianos, G.; Baltikas, V.; Loukovitis, D.; Banti, D.; Sfikas, A.; Papastergiadis, E.; Samaras, P. Biocorrosion of Concrete Sewers in Greece: Current Practices and Challenges. *Sustainability* **2020**, *12*, 2638, doi:10.3390/su12072638.

2. Fytianos, G.; Tziolas, E.; Papastergiadis, E.; Samaras, P. Least Cost Analysis for Biocorrosion Mitigation Strategies in Concrete Sewers. *Sustainability* **2020**, *12*, 4578, doi:10.3390/su12114578.
3. Brad, S.; Murar, M.; Vlad, G.; Brad, E.; Popanton, M. Lifecycle Design of Disruptive SCADA Systems for Waste-Water Treatment Installations. *Sustainability* **2021**, *13*, 4950, doi:10.3390/su13094950.
4. García-López, M.; Melgarejo, J.; Montano, B. The Financing of Wastewater Treatment and the Balance of Payments for Water Services: Evidence from Municipalities in the Region of Valencia. *Sustainability* **2021**, *13*, 5874, doi:10.3390/su13115874.
5. Tsiakiri, E.; Mpougali, A.; Lemonidis, I.; Tzenos, C.; Kalamaras, S.; Kotsopoulos, T.; Samaras, P. Estimation of Energy Recovery Potential from Primary Residues of Four Municipal Wastewater Treatment Plants. *Sustainability* **2021**, *13*, 7198, doi:10.3390/su13137198.
6. Psaltou, S.; Kaprara, E.; Kalaitzidou, K.; Mitrakas, M.; Zouboulis, A. The Effect of Thermal Treatment on the Physicochemical Properties of Minerals Applied to Heterogeneous Catalytic Ozonation. *Sustainability* **2020**, *12*, 503, doi:10.3390/su122410503.
7. Dang, H.T.T.; Dinh, C.V.; Nguyen, K.M.; Tran, N.T.H.; Pham, T.T.; Narbaitz, R.M. Loofah Sponges as Bio-Carriers in a Pilot-Scale Integrated Fixed-Film Activated Sludge System for Municipal Wastewater Treatment. *Sustainability* **2020**, *12*, 4758.
8. Lavrnić, S.; Pereyra, M.Z.; Cristino, S.; Cupido, D.; Lucchese, G.; Pascale, M.; Toscano, A.; Mancini, M. The Potential Role of Hybrid Constructed Wetlands Treating University Wastewater—Experience from Northern Italy. *Sustainability* **2020**, *12*, 604, doi:10.3390/su122410604.
9. Prochaska, C.; Zouboulis, A. A Mini-Review of Urban Wastewater Treatment in Greece: History, Development and Future Challenges. *Sustainability* **2020**, *12*, 6133, doi:10.3390/su12156133.
10. Venieri, D.; Mantzavinos, D.; Binas, V. Solar Photocatalysis for Emerging Micro-Pollutants Abatement and Water Disinfection: A Mini-Review. *Sustainability* **2020**, *12*, 47, doi:10.3390/su122310047.
11. Chung, C.S.; Choi, K.-Y.; Kim, C.-J.; Jung, J.-M.; Chang, Y.S. Overview of the Policies for Phasing Out Ocean Dumping of Sewage Sludge in the Republic of Korea. *Sustainability* **2020**, *12*, 4553, doi:10.3390/su12114553.

 *sustainability*

Article

Biocorrosion of Concrete Sewers in Greece: Current Practices and Challenges

Georgios Fytianos [1,*], Vasilis Baltikas [2], Dimitrios Loukovitis [3,4], Dimitra Banti [1], Athanasios Sfikas [2], Efthimios Papastergiadis [1] and Petros Samaras [1]

[1] Department of Food Science and Technology, International Hellenic University, Sindos, GR-57400 Thessaloniki, Greece; bantidim@gmail.com (D.B.); efspa@food.teithe.gr (E.P.); samaras@food.teithe.gr (P.S.)

[2] DECUS Consultants and Engineers, Ethnikis Antistaseos 4, 55133 Kalamaria, Thessaloniki, Greece; vmpaltik@gmail.com (V.B.); asfikas@decus.gr (A.S.)

[3] Department of Agriculture, International Hellenic University, Sindos, GR-57400 Thessaloniki, Greece; dloukovi@hotmail.com

[4] Research Institute of Animal Science, ELGO Demeter, 58100 Paralimni, Giannitsa, Greece

* Correspondence: gfytianos@gmail.com; Tel.: +30-2310013355

Received: 10 March 2020; Accepted: 23 March 2020; Published: 26 March 2020

Abstract: This paper is intended to review the current practices and challenges regarding the corrosion of the Greek sewer systems with an emphasis on biocorrosion and to provide recommendations to avoid it. The authors followed a holistic approach, which included survey data obtained by local authorities serving more than 50% of the total country's population and validated the survey answers with field measurements and analyses. The exact nature and extent of concrete biocorrosion problems in Greece are presented for the first time. Moreover, the overall condition of the sewer network, the maintenance frequency, and the corrosion prevention techniques used in Greece are also presented. Results from field measurements showed the existence of H_2S in the gaseous phase (i.e., precursor of the H_2SO_4 formation in the sewer) and *acidithiobacillus* bacteria (i.e., biocorrosion causative agent) in the slime, which exists at the interlayer between the concrete wall and the sewage. Biocorrosion seems to mainly affect old concrete networks, and the replacement of the destroyed concrete pipes with new polyvinyl chloride (PVC) ones is currently common practice. However, in most cases, the replacement cost is high, and the authors provide some recommendations to increase the current service life of concrete pipes.

Keywords: sewer corrosion; biocorrosion; concrete sewers

1. Introduction

Sulfide generation is a bacterially mediated process occurring in the submerged portion of sanitary sewage systems from Sulfur-Reducing Bacteria (SRB) [1]. After H_2S diffusion towards the upper part of the sewer pipe above the wastewater, due to the presence of Sulfur-Oxidizing Bacteria (SOB, e.g., *Thiobacillus*), H_2S can be oxidized to biogenic H_2SO_4, which rapidly corrodes the concrete in sewer pipes [1,2]. This oxidizing process can take place wherever there is an adequate supply of H_2S gas (>2 mg/L), high relative humidity, and high atmospheric oxygen content. These conditions are thought to exist in the majority of wastewater systems for at least some times during the year [1]. Figure 1 shows a section of a concrete pipe with the different phases in a typical concrete sewer pipe (adopted by Wu et al. 2018 [2]).

The root cause of biocorrosion is the formation of H_2S, which is produced from sulfates in wastewater under a reaction with sulfate-reducing bacteria located in a slime layer. The slime layer is a layer of bacteria and inert solids at the interface between the concrete wall and the sewage—the submerged portion [3–5]. The slime layer is typically between 0.3 and 1.0 mm thick depending on the flow velocity and solids abrasion in the sewage [6].

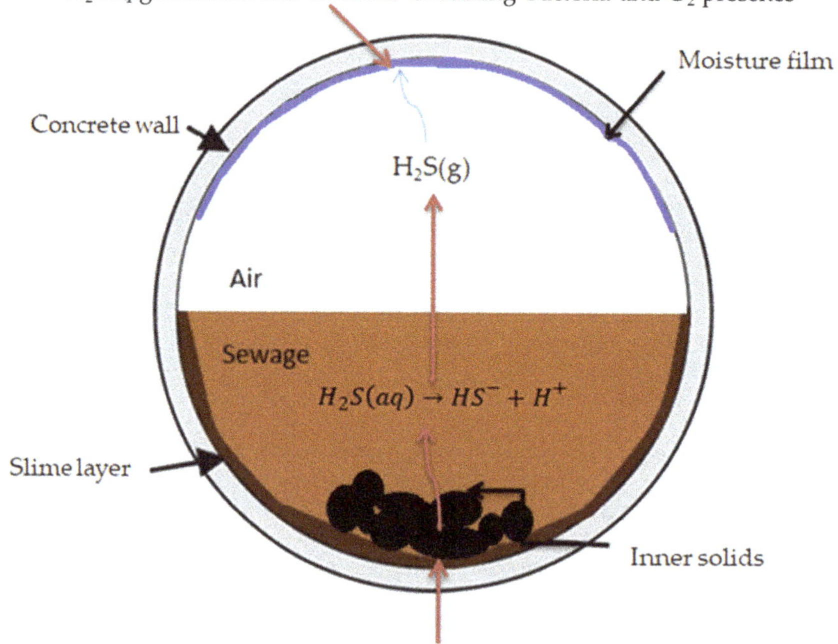

Figure 1. Cross-section representation of a concrete sewer pipe (based on [2]).

As shown in Figure 1, after H$_2$S is generated from sulfates reacting with SRB that are located in the slime layer, it diffuses through the sewage to the air where it can be oxidized to H$_2$SO$_4$ in the presence of SOB. The biogenic H$_2$SO$_4$ then deteriorates the concrete wall.

The basic conditions for the occurrence of biocorrosion are the production of H$_2$S in sewage and the construction of drainage networks from materials that can be corroded by the acids produced by the chemical and biological processes. Biogenic corrosion has been investigated in other European countries. In Flanders, Belgium, biogenic corrosion of sewers costs €5 million annually, representing approximately 10% of the total sewage treatment cost. [7,8].

In Greece, the use of plastic pipes have gradually become common practice since the mid-1980s, beginning with their use in the construction of new drainage systems. However, there are still cement/concrete pipes in operation, especially in the cases of large cross sections and underneath historical places. The sewer network of the two biggest cities of Greece, Athens and Thessaloniki, is made mainly of concrete, while in smaller cities such as Lamia and Komotini, it is made out of PVC.

In addition, the establishment of wastewater treatment facilities in Greece has been on the rise since the 1990s. This means that the probability of occurrence of the phenomenon has increased. The reason is the separation of urban wastewater from industrial wastewater treatment plants, which results in less concentration of heavy metals and chemicals in urban wastewater. Such substances inhibit the growth of the population of microorganisms involved in biocorrosion. The intensity and extent of the phenomenon depends on the configuration and the characteristics of each network separately. In study cases in Greece, the presence of H$_2$S in wastewater is mainly addressed from the point of view of odor management [9], and its treatment seems to have been investigated only with the addition of nitrates (NO_3^-) [10]. With regard to the contribution to scientific research of corrosion-induced concrete drainage pipes, there are publications on the development of mathematical modeling simulations [11,12]. Sulfide can be removed by chemical additives [13,14] or by additives which inhibit biological activity [15], among other methods. Based on available literature and on personal communication with local authorities, it is noted that there is no systematic monitoring and research on biocorrosion in Greece.

An ongoing national R&D project [16] focuses on the development of an innovative active product based on $Mg(OH)_2$ and MgO, for the coating of the inner surfaces of concrete sewer network pipes with corrosion problems. Before moving to the study for the production of the coating, a holistic approach regarding the study of the biocorrosion status in Greece needs to take place.

This paper is intended to review the current practices and challenges of the Greek sewer systems due to biocorrosion and to provide recommendations to avoid it. The authors followed a holistic approach which included survey data obtained by local authorities serving more than 50% of the total country's population, and validation of the survey answers with field measurements and analyses. The objective of this paper is to investigate the extent of corrosion with a special focus on biocorrosion in the Greek sewer network. To do this, authors used a questionnaire as a basic research tool and also conducted field measurements in a representative town experiencing biocorrosion problems.

2. Materials and Methods

A holistic analysis took place with the methodology consisting of two parts. First, the authors wanted to investigate which Greek cities biocorrosion is a valid problem. To this end, a questionnaire, which was answered by 11 local authorities responsible for water and wastewater (i.e. Municipal Water and Sewerage Enterprises, MWSE in this paper, ΔΕΥΑ in Greek), was used. MWSEs are public utilities and one of the major distributors of drinking water in Greece. Contact with bodies and persons related to the operation of sewer networks in Greece was necessary. Second, in order to validate the findings from the questionnaire, representative samples from field measurements (i.e. Kozani) were collected for further tests (i.e., gas analysis, liquid analysis, microscopy of raw solid samples, and molecular genetic analysis of the bacterial slime).

2.1. Questionnaire Development

Based on the knowledge of the authors, no previous study related to the status of corrosion issues in Greek sewer systems existed. Therefore, the use of survey data as a basis research tool similar to the work of [17] was chosen. The survey targeted nine local authorities from different cities as well as the two public companies from the two biggest cities of Greece (EYDAP S.A. from Athens and EYATH S.A. from Thessaloniki) in which 50% of the population resides. Moreover, there are 126 small MWSEs, out of which nine replied to the questionnaire. All respondents were the directors of MSWSEs and had engineering and/or business administration background. Despite the low reply rate from MWSEs, a relatively broad geographical and socioeconomic range was covered. When incomplete or inconsistent data survey data was found, the MWSEs were contacted directly. From the various cases, the authors put special focus in Kozani, a middle-sized town, due to the fact that it showed significant biocorrosion problems based on the results from the questionnaire.

The questionnaire was split into two interconnected parts. The first part included more general questions for the purpose of drawing conclusions regarding:

- the overall state of the sewer network,
- the extent to which the corrosion of sewer pipes is generally recognized as a problem and how it is addressed,
- the extent to which different types of corrosion, especially biocorrosion, are identified as problems with different causes and how they are dealt with,
- the frequency of sewer network inspections and corrosion inhibitions measurements,
- a first estimate of the cost of network maintenance related to corrosion.

The second part of the questionnaire contained questions about specific corrosion incidents, such as the elements of the pipeline where it was found (e.g., material, age, and geometry), the type of corrosion, and repair method. The questionnaire form can be found in Figure 2.

	General Questions
1	**Are you experiencing corrosion problems in the sewer network? What kind of corrosion is the most common?**
	Most common corrosion type: ☐ Yes ☐ No ☐ biochemical ☐ Mechanical ☐ Other:
2	**Describe the corrosion prevention measurements used against corrosion**
	☐ Cleaning ☐ Ventilation ☐ Coating (Material:) ☐ Chemical Additives (chemical name/type:) ☐ Other:
3	**Frequency of the application of corrosion prevention measurements**
	☐ continuously ☐ monthly ☐ trimonthly ☐ semi-annual ☐ annual ☐ never ☐ other:
4	**Inspection frequency of sewer network**
	☐ continuously ☐ monthly ☐ trimonthly ☐ semi-annual ☐ annual ☐ never ☐ other:
5	**Do you keep documentation of the inspections? If yes, what kind of parameters?**
6	**Do you keep documentation related to biocorrosion? Specifically, do you keep documentation on…**
	☐ H_2S concentration ☐ pH ☐ BOD ☐ sewer temperature ☐ sewage temperature ☐ flow rate ☐ sewage velocity ☐ other: ☐ Nothing related
7	**What is the total maintenance cost of the sewer network?**
8	**What is the replacement cost of a sewer pipe? (euro/m)**
9	**What is the total cost of replacement/maintenance due to corrosion?**
10	**What is the cost for the application of corrosion prevention measurements? (if any)**
11	**Please indicate the type of the sewer system (combined or separated), material, length and age**
12	**Please write any additional comments**

Figure 2. Questionnaire form of the survey.

The questionnaire results were based on personal estimations of the MWSEs directors rather than on quantitative data. Therefore, to validate the findings of the questionnaire, sample analysis was necessary for examining the existence of H_2S and the existence of biogenic sulfide corrosion bacteria in sewer pipes (Sections 2.2 and 2.3).

2.2. Analysis of Samples

Field measurements in Kozani were carried out to quantify H_2S, CH_4, and O_2 in the gaseous phase of the sewers, as well as to take solid and liquid samples from the pipes. For the solids and liquids, sterile containers were used for the transportation of the samples to the laboratory.

H_2S, CH_4, and O_2 gas measurements of the sewer network were conducted with a validated Eurotron Rasi700 Bio automated portable analyzer. The unit was configured to measure H_2S, and with the help of the local employees, the manhole cover was opened and the gas measurement took place swiftly. The tube of the analyzer was stopped at a depth close to the top of the sewers (Figure 3a). Measurements were made under normal operating conditions (i.e., no clogging or blockage) of the sewer network. Liquid analysis of the sewage for Chemical Oxygen Demand (COD), Total Organic Carbon (TOC), total Nitrogen, and Total Phosphorus was done with the use of Merck test kits.

(a)

(b)

Figure 3. Field test in Kozani (**a**) Measurement for H_2S, CH_4, O_2 gases; (**b**) Sample collection from sewer.

Raw samples of sewer pipes and the materials on them were collected (Figure 3b) and examined microscopically with a Carl Zeiss™ Stemi 2000-C Stereo Microscope. It should be noted that most samples were heavily deteriorated and degraded due to corrosion.

2.3. Molecular Genetic Analysis of Bacterial Slime

A semi-solid sample of sludge, coming from a pipe (slime layer) of the sewer network was used for bacterial community analysis. 300 mg of sample material were used for genomic DNA extraction with the 'NucleoSpin Soil' kit (Macherey-Nagel, Germany) following the manufacturer's protocol. The quality and quantity of the isolated DNA were checked with a ND-2000 NanoDrop Spectrophotometer (Thermo Fisher Scientific, USA) and by electrophoresis on a 1% agarose gel, stained with Midori Green DNA stain (NIPPON Genetics Europe, Germany).

50 µl of DNA sample, with a concentration of 250 ng/µl, were sent to CeMIA S.A. (Greece) for 16S rRNA-based microbial profiling. Nowadays, 16S metagenomics is considered to be one of the most reliable methods for microbial diversity analysis of mixed samples by utilizing next generation sequencing technology. Combined with proper bioinformatic analysis, taxonomic classification of microbes is performed down to family/genus level, while in some cases, species-level resolution can be achieved. Various hypervariable regions of the bacterial 16S rRNA gene (V2, V3, V4, V6-7, V8, and V9) were amplified with two sets of primers using the Ion 16S™ Metagenomics Kit (ThermoFisher Scientific, USA). The amplified fragments were then sequenced on the Ion Torrent S5XL platform (ThermoFisher Scientific, USA) and analyzed using the Ion 16S™ metagenomics analyses module within the Ion Reporter™ software (https://ionreporter.thermofisher.com/ir/).

3. Results

3.1. Survey Results

Eleven questionnaires were returned from MWSEs. The location of the towns that replied can be found in Figure 4. On the basis of the answers provided, data was collected from the sewer networks of 12,000 km total length, serving 50% of the population of Greece. In Table 1, a detailed list of the cities and answers given is presented. The town numbers of Table 1 correspond to the numbers in Figure 4. The "Peak Population Equivalent" data was obtained by the monitoring database of the special secretariat for water, Ministry of Environment and Energy [18]. In Table 1, a combined system carries both surface run-off and wastewater, while a separate system carries the municipal wastewater and surface run-off separately.

Figure 4. Map with the cities which answered the survey (map taken from [19]).

Altogether, eight of the MWSEs responded that they encountered pipeline corrosion problems, while three did not. The three MWSEs citing no corrosion refers only to small provincial MWSEs (i.e., Lamia, Komotini, Tyrnavos) and is explained by the fact that these networks are relatively new (after 1990). In addition, most of the non-corroded sewer parts at these three MWSE are made from PVC.

Table 1. Details of Sewerage Enterprises which were answered in the questionnaire.

Town	Town Number on Map	Peak Population Equivalent	Type of System	Corrosion Problems	Type of Corrosion	Corrosion Prevention Measurement	Inspection Frequency	Prevention Measures Frequency
Athens	1	5200000	Combined	Yes	mainly Mechanical	Cleaning, Ventilation	continuously	Semi-annual
Thessaloniki	2	900000	Combined	Yes	Biochemical/Mechanical	Cleaning, Ventilation	Semi-annual	Semi-annual
Ioannina	3	142000	Separate	Yes	Biochemical/Mechanical	Cleaning	trimontly	Trimontly
Serres	4	79000	Combined	Yes	Biochemical/Mechanical	Cleaning, Ventilation	montly	Montly, n.s.
Lamia	5	78200	Separate	No	n/a	Cleaning	trimontly	n/a
Komotini	6	72000	Separate	No	n/a	Cleaning	Annual, n.s.	n/a
Kozani	7	46000	Combined	Yes	Biochemical/Mechanical	Cleaning, Chemical Additives	montly	Montly
Agios Nikolaos	8	25000	Separate	Yes	Biochemical	Cleaning, Other	Annual	Annual
Florina	9	20000	Separate	Yes	Biochemical	Cleaning, Ventilation	montly	Semi-annual
Tyrnavos	10	10900	Separate	No	n/a	Cleaning	trimontly, n.s.	n/a
Chortiatis	11	4800	Combined	Yes	Mechanical	Cleaning, Other	montly	Semi-annual

According to the results of those who responded positively to the occurrence of corrosion in the networks they manage, the matter of which type of corrosion is most commonly encountered arose. Thus, for all respondents, the following results are shown: four biochemical and mechanical corrosion, two only biochemical corrosion, two only mechanical corrosion, and three no corrosion problems.

It seems that the corrosion phenomenon of the sewer pipelines of the Greek sewerage systems is a problem according to the replies of the majority of the respondents. In general, the use of all types of plastic pipelines in Greece began in the late-1980s. Therefore, the absence of corrosion problems has a reasonable basis in this respect.

The fact that the highest percentage (37%) of the types of corrosion found belongs to the combination of biochemical and mechanical corrosion mechanisms reflects the fact that chemical and mechanical corrosion can occur in combination.

On the question of whether corrosion prevention measures are implemented, all respondents answered that the drains were cleaned, four were ventilated, one added chemical additives to the sewage, and three have done something else. In the category "Other", some MWSEs use the addition of microorganisms for fat removal, but no company uses coatings onto the inner surface of the pipelines.

The removal of solids is also of great importance in reducing mechanical corrosion. According to the answers given, the ventilation of the network is mainly for deodorizing purposes. However, it results in a decrease in the concentration of H_2S in the atmosphere of the pipelines, which also reduces the formation of H_2SO_4 and thus, corrosion. Again, mainly for deodorizing purposes, some MWSEs have added chemical additives to the sewage.

MWSEs were also asked about the inspection frequency of the sewer pipes under their responsibility and for the frequency of implementation of preventive measures, namely whether they are approximately monthly, trimonthly, semi-annual, or annual. Not all MWSEs inspect the sewer network monthly. Specifically, to the question of when the sewer pipes under their responsibility are inspected, four answered every month, three trimonthly, one semi-annual to annual, while two of the respondents answered that they inspect them approximately once a year. By weighting the above answers based on their frequency, it appears that in the Greek territory, the condition of sewer networks is inspected on average about six times a year and that the precautionary measures are applied approximately four times a year.

MWSEs were asked if they kept records of the inspections of the networks and what type of records they contained. They were also asked if they had been measuring and recording H_2S concentrations, pH, COD, sewage temperatures, effluents properties, and flow rates. These are all parameters that, together with the geometrical characteristics of the network, can be used for calculations of hydrogen sulfide production, risk and/or corrosion rate. Unfortunately, such records are not systematically maintained for the sewer networks. These kinds of analyses are carried out at the inputs and output of the wastewater treatment plants, and only on some parameters of the incoming sewage and the effluent of their treatment.

It is very difficult to determine the cost of repairs for sewer pipes due to damage caused by corrosion. None of the wastewater and sanitation enterprises involved in the survey calculates this separately. The pipes are usually replaced when they are seriously deteriorated. The information provided are business estimates of the total network maintenance costs and cost per meter for pipeline replacement. The average maintenance cost of a sewer network in Greece was calculated based on the total costs and network length of each enterprise. The average cost of pipeline replacement was calculated accordingly, and the average cost of pipeline replacement presents business-to-business variations, depending on the extent to which the work is performed by the same resources or by third parties through project contracts. A more detailed determination of the costs specifically associated with corrosion was not carried out in this study. The authors' estimation based on the answer from the questionnaire is 375 €/km as the average maintenance cost, and 200 €/m as average replacement cost (without including the salaries of external contractors).

In Thessaloniki, the sewer network is 35% combined [20]. It is noteworthy that the main part of the sewerage system of the city dates back to 1926 and was constructed in the context of the rehabilitation of the city after the devastating fire of 1917. A disadvantage of the city's sewer network is the lack of its ventilation infrastructure which favors the occurrence of corrosion due to the presence of hydrogen sulfide. According to the local authorities, about 90% of the maintenances related to the sewer network concern concrete damages (reinforced and not).

Figure 5 shows a part of the old sewage network. The lower part of the pipe is covered with ceramic tiles. Deposits of fat and possibly sulfur compounds are observed. The upper surface shows some corrosion, although it appears to be progressing slowly with respect to the age of the pipe, which may be as old as 93 years. Potential fat deposits on the sewer pipes may enhance anaerobic conditions which favor the production of H_2S, and hence biocorrosion. Furthermore, high content of fats results in pipe blockage.

Figure 5. Main sewage pipeline of Thessaloniki.

However, there are numerous cases of pipe corrosion issues. In the following diagram (Figure 6), the corrosion-related damages distribution for 2016–2018 by the decade of construction of pipelines is presented. It should be noted that the majority of damages concern concrete pipes constructed until the 1970s with 83%, while 14% concern pipes up to 40 years since construction.

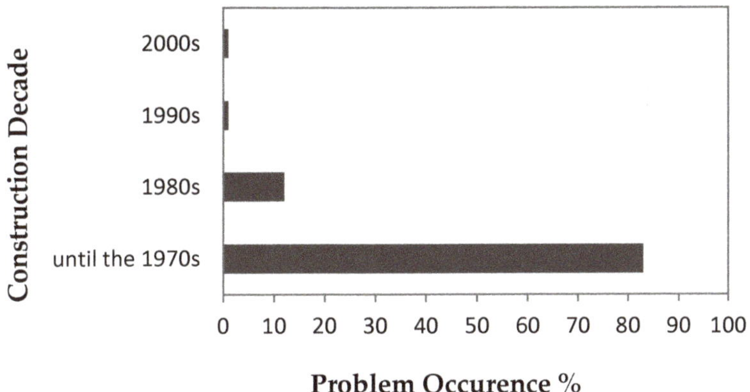

Figure 6. Pipe damage occurrence distribution in Thessaloniki's sewer system for the years 2016–2018 based on the age of the pipe.

In Athens, pipeline corrosion can mainly be characterized as mechanical, which in some cases is secondarily affected by H_2S chemical corrosion (based on the questionnaire answers). Corrosion of these pipes is observed in very old combine system pipes.

The sewerage network is maintained and constantly checked to avert problems that create damage to the roadway and to minimize any malfunctions. In order to locate and repair damages in the sewerage network, the responsible authority uses mobile units that inspect the network telescopically. These mobile units contain the recording, photographic, and video-recording systems that convey all the data that the camera collects from inside the pipe (such as the exact location and nature of the damage) to a computer. The cameras are used to inspect pipes ranging in diameter from 200 to 1500 mm, as well as for the inspection of individual building connections. For pipe sections of a larger diameter, which can accommodate direct inspection by technical personnel, the cameras can be adjusted to a portable system. With the use of that technology, lower maintenance costs and quicker repair time are achieved while minimizing social annoyance from unnecessary digging [21].

The problems of corrosion of sewer pipes of the MWSE of Kozani network are found in the city's combined sewage system. The first pipes were installed in the 1950s, and in the mid-1980s, they were replaced with newer ones from the same construction material. The networks consist of pipes 1 m long and there are some problems in the connections between them. Based on personal communication, there have been numerous cases of slime and bad odor in the sewers. Some parts of the network are completely destroyed due to corrosion. Based on the microbiological results and the questionnaire, it seems that Kozani has experienced serious biocorrosion problems in the sewer network. It was reported by the local authorities that corrosion failure is found not only in the upper part of the pipeline, but also in the lower part and on the sides. In addition to that, the slime is found under the pipeline close to the connections between the pipes (due to the improper connection). In Figure 7, MWSE staff is dealing with pipeline failure, and in Figure 8, the cement pipe is completely degraded and only the slime is visible.

Figure 7. Corrosion failure in Kozani's sewage pipe.

Figure 8. Total degradation of cement pipe due to biocorrosion in Kozani.

The strong point of the methodological approach used in this study was the combination of survey results with experimental data. Similar studies [2] for the city of Edmonton focused on hydraulic parameters and sewer system design, which were not investigated in this study.

3.2. Field Measurements Results

3.2.1. Gas Analysis

The results for H_2S, CH_4, and O_2 are presented in Table 2 and in Figure 9. The authors chose Kozani as a study case, because from the questionnaire results, it seemed that biocorrosion was the corrosion type. In most of the test sites, small amounts of H_2S were found, which could indicate possible biocorrosion. It should be noted that during tests, the sewerage system was under normal operating conditions and no clogs were observed. 2 ppm (mg/L) of H_2S is a sufficient concentration to lead to biocorrosion. Furthermore, H_2S concentrations ranging from 2 to 5 ppm may cause nausea and headaches, while concentrations from 100 ppm can cause coughing, throat irritation, and death.

Table 2. Gas measurements during the field test in Kozani.

Place	Measured Gas		
	H_2S (ppm)	CH_4 (%)	O_2 (%)
1	2	0.04	21
2	1	0.03	20.3
3	1	0	20.9
4	0	0.08	21.1
5	0	0005	21
6	1	0.16	20.9
7	1	0.03	20.9
8	2	0.03	20.5
9	2	0.2	20.6
10	1	0.09	20.8
11	1	0.04	20.4
12	1	0.03	20.2
13	1	0.02	20.9

Map Location	Coordinates
1	40.298518,21.788187
2	40.298707,21.788015
3	40.297645,21.788206
4	40.297928,21.786011
5	40.300662,21.790603
6	40.300558,21.790611
7	40.289420,21.781500
8	40.288392,21.781248
9	40.294618,21.783009
10	40.297247,21.782826
11	40.297252,21.782804
12	40.297983,21.783493
13	40.297939,21.783997

Figure 9. Sampling points corresponding to Table 2.

3.2.2. Liquid Analysis

In Table 3, average results of three samples for COD, TOC, total nitrogen, and total phosphorus from the sewage collected from Kozani are presented. As shown in Table 3, the values are matching typical values for urban wastewater [22].

Table 3. Liquid Analysis Results from the sewage sample.

Sewage Chemical Parameters	mg/L
Chemical Oxygen Demand (COD)	910
Total Organic Carbon (TOC)	238,7
Total Nitrogen	49
Total Phosphorus	5,1

3.2.3. Concrete Analysis

Images taken with the microscope are presented in Figure 10. A set of four pictures with increasing magnification are shown. What appears to have solidified onto the surface is the bacteria slime (it can be observed in the latter two images with higher magnification).

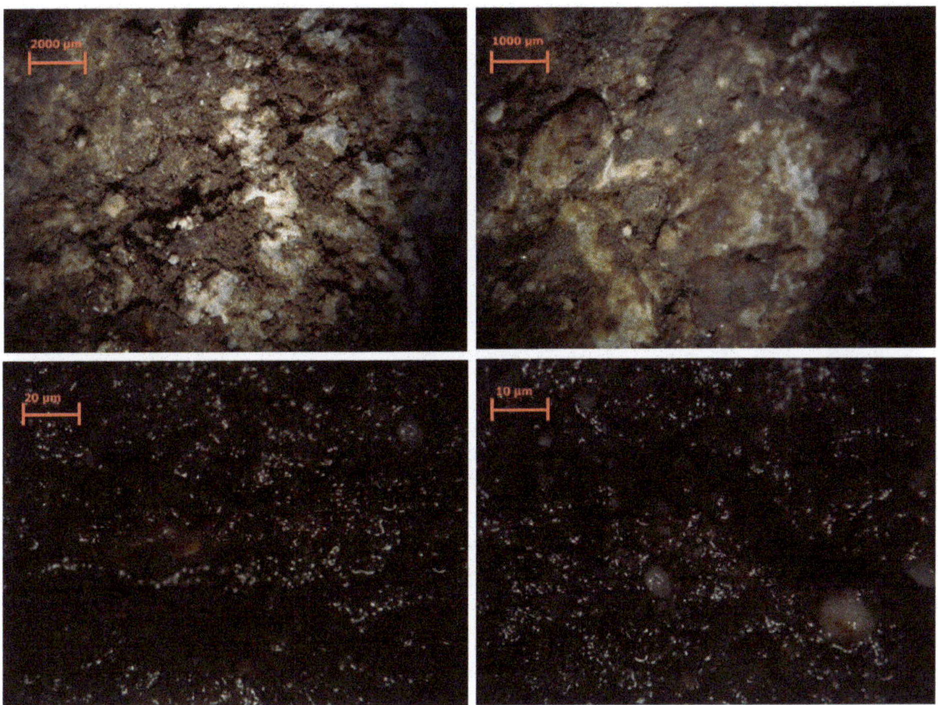

Figure 10. Microscopic images with different magnitudes of a corroded sample from Kozani.

3.2.4. Slime Genetic Analysis

The microbial diversity analysis of the DNA sample revealed a wide spectrum of bacterial species that were present in the slime layer. The resolution of the microbial profiling was, in most cases, feasible down to family/genus level, and in some cases, down to species level. Bacteria belonging to various phyla were identified, i.e., *Acidobacteria, Actinobacteria, Bacteroidetes, Chlamydiae, Chloroflexi, Cyanobacteria, Firmicutes, Gemmatimonadetes, Ignavibacteriae, Nitrospirae, Planctomycetes,* and *Proteobacteria*. Among the bacterial families detected was also the *Acidithiobacillaceae* family (order: *Acidithiobacillales,* class: *Gammaproteobacteria,* phylum: *Proteobacteria*). The specific family contains a single genus, *Acidithiobacillus,* with *Acidithiobacillus thiooxidans* as the type species. Four other species of this genus are currently recognized: *At. ferrooxidans, At. caldus, At. albertensis,* and *At. ferrivorans.* The *Acidithiobacillus* genus is of special interest because its species include some of the most extremely acidophilic bacteria known, which tolerate extraordinarily high concentrations of some toxic metals. *Acidithiobacillus thiooxidans* oxidizes sulfur and produces sulfuric acid, and it has also been observed, causing biogenic sulfide corrosion of concrete sewer pipes by altering hydrogen sulfide in sewage gas into sulfuric acid [23].

4. Recommendations

Some recommendations to mitigate biocorrosion in concrete sewers are as follows:

- The regular measurement and recording of H_2S concentrations, pH, COD, sewage temperatures, effluents properties, and flow rates. These are all parameters that, together with the geometrical characteristics of the network, can be used for calculations of hydrogen sulfide production and risk.
- Use of mobile units equipped with cameras for regular inspections.
- Surface washes with water. Although flushing with high-pressure water removes the corrosion deposits from the concrete surface and increases the surface pH, the effects are short term, i.e., one month [24], or two to four months [25] and for a long-term protection, frequent flushing with high-pressure water is necessary.
- Treatment of the concrete surface so as to be less susceptible to corrosion. This can be done by using spray-on coatings, e.g., $Mg(OH)_2$ based coatings.
- Application of polyethylene (PE) liner.
- Inhibition of the biological activity, e.g., with biocides.

For proper concrete sewer system design, avoiding sedimentation in sewer conduits should be taken into account. Towards this direction, mathematical modeling could be beneficial [26].

Each case is different, and a life cycle costing analysis for each method could be advantageous in order to estimate the most cost-efficient biocorrosion mitigation methodology.

5. Conclusions

The results from the questionnaire showed that corrosion is present in Greece's sewer networks and has caused the destruction of sewer pipe sections made of concrete. The replacement of the destroyed concrete pipes with new polyvinyl chloride (PVC) ones is currently common practice. Further gas and slime genetic analysis supported the findings of the questionnaire and showed that in the case of Kozani, biocorrosion is the main type of corrosion that takes place. Biocorrosion seems to affect mainly old networks, city centers, and large diameter collectors. As a next stage, since most of the concrete networks cannot be replaced easily and economically, the authors will examine the effectiveness of a protective coating based on $Mg(OH)_2$ and MgO that can be applied onto the concrete surfaces as a solution to control biocorrosion. For future studies, Life Cycle Cost Analysis (LCCA) can be a useful tool for the economic evaluation of various biocorrosion mitigation strategies.

Author Contributions: Conceptualization, G.F. and P.S.; Methodology, G.F., A.S., and P.S.; Software, G.F., D.L., E.P. and V.B.; Validation, G.F., D.B. and V.B.; Formal Analysis, G.F. and D.L.; Investigation, G.F. and V.B.; Resources, P.S.; Data Curation, G.F.; Writing—Original Draft Preparation, G.F., V.B.; Writing—Review and Editing, D.B., P.S.; Visualization, G.F. and E.P.; Supervision, P.S.; Project Administration, P.S.; Funding Acquisition, P.S. and A.S. All authors have read and agreed to the published version of the manuscript.

Funding: This research has been co-financed by the European Union and Greek national funds through the Operational Program Competitiveness, Entrepreneurship and Innovation, under the call RESEARCH—CREATE—INNOVATE (project code:T1EDK-02355-title 'Novel Coating Materials for Corrosion Protection of Sewer Network Pipes'.)

Conflicts of Interest: The authors declare no conflict of interest.

References

1. O'Dea, V. Understanding biogenic sulfide corrosion. *Mater. Perform* **2007**, *46*, 36–39.
2. Wu, L.; Hu, C.; Liu, W.V. The sustainability of concrete in sewer tunnel—A narrative review of acid corrosion in the city of Edmonton, Canada. *Sustainability* **2018**, *10*, 517. [CrossRef]
3. Yamanaka, T.; Aso, I.; Togashi, S.; Tanigawa, M.; Shoji, K.; Watanabe, T.; Watanabe, N.; Maki, K.; Suzuki, H. Corrosion by bacteria of concrete in sewerage systems and inhibitory effects of formates on their growth. *Water Res.* **2002**, *36*, 2636–2642. [CrossRef]
4. Mori, T.; Nonaka, T.; Tazaki, K.; Koga, M.; Hikosaka, Y.; Noda, S. Interactions of nutrients, moisture, and pH on microbial corrosion of concrete sewer pipes. *Water Res.* **1992**, *26*, 29–37. [CrossRef]
5. Okun, D.A.; Wang, L.K.; Shammas, N.K. *Water Supply and Distribution and Wastewater Collection*; JohnWiley and Sons: Hoboken, NJ, USA, 2010.

6. Bowker, R.P.; Smith, J.M. Odor and Corrosion Control in Sanitary Sewerage Systems and Treatment Plants; U.S. Environmental Protection Agency: Washington, DC, USA, 198 e sewer pipes. *Water Res.* **1992**, *26*, 29–37.

7. Vincke, E. Biogenic Sulfuric Acid Corrosion of Concrete: Microbial Interaction, Simulation and Prevention. Ph.D. Thesis, Faculty of Bio-engineering Science, University Ghent, Ghent, Belgium, 2002; pp. 7–9.

8. Zhang, L.; De Schryver, P.; De Gusseme, B.; De Muynck, W.; Boon, N.; Verstraete, W. Chemical and biological technologies for hydrogen sulfide emission control in sewer systems: A review. *Water Res.* **2008**, *42*, 1–12. [CrossRef] [PubMed]

9. Karageorgos, P.; Latos, M.; Kotsifaki, C.; Lazaridis, M.; Kalogerakis, N. Treatment of unpleasant odors in municipal wastewater treatment plants. *Water Sci. Technol. J. Int. Assoc. Water Pollut. Res.* **2010**, *61*, 2635–2644. [CrossRef] [PubMed]

10. Mathioudakis, L.V.; Vaiopoulou, E. Addition of nitrate for odor control in sewer networks: Laboratory and field experiments. *Glob. NEST J.* **2006**, *8*, 37–42.

11. Nikolopoulos, C.V. A mushy region in concrete corrosion. *Appl. Math. Model.* **2010**, *34*, 4012–4030. [CrossRef]

12. Nikolopoulos, C.V. Macroscopic models for a mushy region in concrete corrosion. *J. Eng. Math.* **2015**, *91*, 143–163. [CrossRef]

13. Henze, M.; Comeau, Y. *Wastewater characterization. Biological Wastewater Treatment: Principles, Modelling and Design*; IWA Publishing: London, UK, 2008; Volume 7, pp. 33–52.

14. Nielsen, A.H.; Lens, P.; Vollertsen, J.; Hvitved-Jacobsen, T. Sulfide–iron interactions in domestic wastewater from a gravity sewer. *Water Res.* **2005**, *39*, 2747–2755.

15. Jayaraman, A.; Mansfeld, F.B.; Wood, T.K. Inhibiting sulfate-reducing bacteria in biofilms by expressing the antimicrobialpeptides indolicidin and bactenecin. *J. Ind. Microbiol. Bio Technol.* **1999**, *22*, 167–175. [CrossRef]

16. Crownmag. Innovative Coating Materials for Corrosion Protection of Sewer Network Pipes. Available online: http://crownmag.gr/en/ (accessed on 10 January 2020).

17. Dorji, U.; Tenzin, U.M.; Dorji, P.; Wangchuk, U.; Tshering, G.; Dorji, C.; Shon, H.; Nyarko, K.B.; Phuntsho, S. Wastewater management in urban Bhutan: Assessing the current practices and challenges. *Process Saf. Environ. Prot.* **2019**, *132*, 82–93. [CrossRef]

18. Wastewater Treatment Plants. Monitoring Database. Special Secretariat for Water. Ministry of Environment and Energy. Available online: http://astikalimata.ypeka.gr/Services/Pages/WtpViewApp.aspx (accessed on 10 January 2020).

19. Available online: http://www.maps-of-europe.net/maps-of-greece/ (accessed on 22 January 2020).

20. Available online: https://www.eyath.gr/sewerage/?lang=en (accessed on 24 January 2020).

21. Available online: https://www.eydap.gr/en/TheCompany/DrainageAndSewerage/NetworkMaintenance/ (accessed on 24 January 2020).

22. Metcalf & Eddy, Inc. *Wastewater Engineering: Treatment and Reuse*; McGraw-Hill: Boston, MA, USA, 2003.

23. Kelly, D.P.; Wood, A.P. The Family Acidithiobacillaceae. In *The Prokaryotes*; Rosenberg, E., DeLong, E.F., Lory, S., Stackebrandt, E., Thompson, F., Eds.; Springer: Berlin/Heidelberg, Germany, 2014.

24. Nielsen, A.H.; Vollertsen, J.; Jensen, H.S.; Wium-Andersen, T.; Hvitved-Jacobsen, T. Influence of pipe material and surfaces on sulfide related odor and corrosion in sewers. *Water Res.* **2008**, *42*, 4206–4214. [CrossRef] [PubMed]

25. Sun, X.; Jiang, G.; Chiu, T.H.; Zhou, M.; Keller, J.; Bond, P.L. Effects of surface washing on the mitigation of concrete corrosion under sewer conditions. *Cem. Concr. Compos.* **2016**, *68*, 88–95. [CrossRef]

26. Song, Y.H.; Yun, R.; Lee, E.H.; Lee, J.H. Predicting sedimentation in urban sewer conduits. *Water* **2018**, *10*, 462. [CrossRef]

 *sustainability*

Article

Least Cost Analysis for Biocorrosion Mitigation Strategies in Concrete Sewers

Georgios Fytianos [1],* , Emmanouil Tziolas [2], Efthimios Papastergiadis [1] and Petros Samaras [1]

[1]   Department of Food Science and Technology, International Hellenic University, Sindos, 57400 Thessaloniki, Greece; efspa@food.teithe.gr (E.P.); samaras@food.teithe.gr (P.S.)

[2]   ELGO-Demeter, Fisheries Research Institute, 64007 Nea Peramos, Kavála, Greece; tziolasm@inale.gr

*   Correspondence: gfytianos@gmail.com; Tel.: +30-2310013355

Received: 15 May 2020; Accepted: 2 June 2020; Published: 4 June 2020

Abstract: The changing role of the municipal water and wastewater authorities, together with the need for a sustainable maintenance treatment in the sewer systems, have been the catalysts for the integration of technical and financial information into asset management systems. This paper presents results from a cost-comparative analysis focusing on an annuities calculation for the evaluation of microbiologically induced corrosion (MIC) or biocorrosion mitigation methodologies used in the maintenance of concrete sewers. The replacement cost of deteriorated sewer concrete pipes is high, and MIC mitigation methods can be used to increase the current service life of concrete pipes. From the MIC mitigation methods that are frequently used, the authors examined those of flushing with high-pressure water (i.e., a common method used in Greece), and spraying with magnesium hydroxide slurry (MHS). The authors chose four different cities for the assessment, which presented different sewer characteristics and socioeconomic backgrounds. In addition, all methods for concrete sewer MIC mitigation were compared to the present value of replacement of sewer concrete pipes with new PVC ones. Results showed that flushing with high-pressure water is very cost demanding and should be avoided, while spraying with MHS could be a sustainable and economic solution in the long term.

Keywords: sewer corrosion; least cost analysis; biocorrosion

1. Introduction

Microbiologically induced corrosion (MIC) or biocorrosion in concrete sewers is a known problem [1]. The root cause of MIC is the formation of H_2S, which is produced from sulfates in wastewater under a reaction with sulfate-reducing bacteria located in a slime layer. H_2S can be oxidized to H_2SO_4, due to the presence of sulfur-oxidizing bacteria, which rapidly corrodes the concrete in sewer pipes. In a concrete sewer wherever there is an adequate H_2S gas (>2 mg/L), high relative humidity, and high atmospheric oxygen content, MIC can take place. These conditions are thought to exist in the majority of wastewater systems for at least some times during the year [2]. Worldwide, the economic losses due to MIC are very high [3]. Annual rehabilitation costs were estimated to reach over USD 400 million in Germany and over USD 100 million in the United Kingdom [4]. In Flanders, Belgium, biocorrosion related costs are approximately 10% of the total sewage treatment cost [5]. Worldwide, it is difficult for municipalities to repair old and deteriorated concrete sewers since sometimes sewers are under historic and populated places [6].

Concrete is still the most used material for sewer pipes in Greece, while for newly installed sewer systems, PVC pipes are favored due to their better corrosion resistance properties and lower price. In Greece there is no systematic monitoring and research on biocorrosion, and based on a recent survey, the local directors of sewerage treatment enterprises are willing to try new MIC mitigation strategies [7].

So far, in study cases in Greece the presence of H_2S in sewers has mainly been addressed with a focus on odor management [8].

Several studies have implemented a life cycle assessment (LCA) methodological approach to assess sewage treatment plants [9,10] and wastewater treatment plants [11,12], while the perspective of these assessments is biased towards the treatment's operational stage. Many related studies take into account the construction phase, apart from the operational stage, which is responsible for only 20% or less of the total environmental impacts [13], though a significant number of studies dwell on the pipe materials [14,15]. Nevertheless, in the abovementioned assessments of sewage and wastewater treatment plants, the maintenance and repair aspect of deteriorating concrete is neglected. The poor durability of these constructions will enhance negative impacts to human health and the infrastructure, while time will spread the impacts to several parts of the infrastructures and aggravate the plants' stability [16] with significant economic impacts [17].

Regarding the pure economic cost of sewer rehabilitation and/or repair-maintenance alternatives, several aspects (e.g., discounting factor, inflation rate, investment period etc.) could create significant trade-offs among the alternatives. More specifically, Æsøy et al. [17] have stated the immense cost fluctuation of different projects across the whole world caused by different intervention strategies and different periods of time. Furthermore, a more recent study illustrates that the replacement cost for a relevant project accounts for more than 50% of the total investment cost [18], highlighting the need for a cost-efficient long-term strategy in a centralized system. The evaluation of alternatives could be done based on an annuities calculation and the depiction of costs in the form of a net present value (NPV), following a solid perspective for the right choice in the rehabilitation of sewers [19]. The operations of treatment and disposal of sewage sludge accounts for approximately 40% to 50% of the total operational costs of a wastewater treatment plant [20]. Focusing on the distribution of costs in wastewater systems, the maintenance aspect accounts for more than one fourth of the total costs [21], whilst specifically in concrete pipes the costs are dependent on the interaction between the discount and inflation rate [22]. Therefore, there is a significant cost connected to the repair and maintenance actions taken. In this context, the management of operational actions related to sewer networks is multi-faceted and requires the investigation of economic and environmental consequences as well [23].

Focusing on Greece, the changing role of the local Municipal Water and Sewerage Enterprises (MWSE in this paper, ΔΕΥΑ in Greek) together with the need for a sustainable maintenance treatment in the sewer systems, has been the catalyst for the integration of technical and financial information into asset management systems. Thus, the application of least cost analysis to sewer installation and maintenance projects has increased in recent years [24]. MWSEs, due to the limitations in their budget, need to include some type of financial analysis for their corrosion maintenance process.

Economic data is more tangible, and can be exported at any time with ease through various methods of qualitative and quantitative research. Unlike information related to air pollutant emissions, that require ideal conditions and expensive equipment to measure, financial data is more easily accessible and can also be targeted to selected areas. The development of a financial assessment for the investigation of economic impacts related to corrosion controlling methods will formulate the main perspective of the present research. In this context, a focus on the design and management of maintenance for the existing structures is given [25], while calculating the manifold impacts for every corrosion controlling method. To further expand upon this, there is a necessity for the economic assessment of the current sewer system in Thessaloniki, Serres, Kozani, and Florina based on the survey of Fytianos et al. [7]. Thessaloniki, as a capital of Northern Greece, should have higher social impacts due to maintenance in the city center, while Kozani and Florina should have difficulties with relevance to the cleaning process due to the smaller pipe sizes. Therefore, given the set boundaries, the economic aspects for each controlling method in the four cities are highlighted, since the local characteristics should be carefully considered because they could affect the total annual costs [26]. In addition, the corrosion mitigation methods' net present values costs are compared to the replacement cost of sewer concrete pipes with new PVC ones.

The objective of this paper is to investigate the future cost of MIC mitigation strategies and to provide the optimized solution for each case. To this end, the authors developed a cost-comparative framework, based on net present value (NPV), as a basic research tool. The different mitigation strategies can be categorized into three groups [3,27,28]; (i) proper sewer hydraulic design, (ii) concrete performance improvement, (iii) and altering sewer pipe environment. This paper focuses on the mitigation strategies for the already installed concrete sewers. Therefore, the MIC mitigation strategies that will be examined focused on altering the sewer pipe environment i.e., controlling the sulfide production.

2. Materials and Methods

2.1. MIC Mitigation Methods

From the MIC mitigation methods that are frequently used, the authors examined those of flushing with high-pressure water (i.e., a common method used in Greece) and spraying with magnesium hydroxide slurry (MHS). MHS has been used successfully for the corrosion control of concrete sewers [29].

2.1.1. Flushing with High-Pressure Water

The use of high-pressure water for flushing the concrete surface has been studied [30,31]. Although flushing with high-pressure water removes the corrosion deposits from the concrete surface and increases the surface pH, the effects are short term, i.e., from 2 weeks [30] to 2–4 months [31], and for long term protection, frequent flushing with high-pressure water is necessary. In the current work, the authors used a monthly frequency scenario for high-pressure water flushing of concrete pipes.

2.1.2. MHS Spraying

Compared to other chemical treatment methods, spraying with MHS is considered to be a very cost-effective methodology [29], which can add additional years of service life to a concrete sewer. MHS will raise the concrete wall's pH to around 9.5. If an adhesion additive is used in the mix, since MHS has very low solubility in water, it can stay on the concrete for sufficient time. An ongoing national R&D project led by Grecian Magnesite S.A. [32] focuses on the development of a product, based on $Mg(OH)_2$ and MgO, for the coating of the inner surfaces of concrete sewer network pipes with corrosion problems. In this paper, the authors examined the case of one spraying of MHS per year. For the application area of a magnesium hydroxide coating (MHC) by spraying, the area of the inner concrete pipe is calculated as:

$$\text{Pipe Surface} = 2 \, \pi \, r \, h \tag{1}$$

where r is the radius of each sewer pipe and h is the length. The volume of MHC that would be needed can be calculated as:

$$V_{MHC} \, [m^3] = \text{Pipe surface} \, [m^2] \times \text{MHC thickness} \, [m] \tag{2}$$

where MHC thickness is estimated at 0.005 m. The mass of the needed MHC is calculated as:

$$m_{MHC} \, [kg] = \rho_{MHC} \, [kg/m^3] \times V_{MHC} \, [m^3] \tag{3}$$

where $\rho_{MHC} = 1500 \, kg/m^3$.

For example, (Figure 1) for a pipe of 1 m length and 1 m diameter the pipe surface would be equal to 3.141 m^2, and from Equation (2), the V_{MHC} that would be needed is equal to 0.0157 m^3. The corresponding m_{MHC}, from Equation (3) is equal to 23.55 kg.

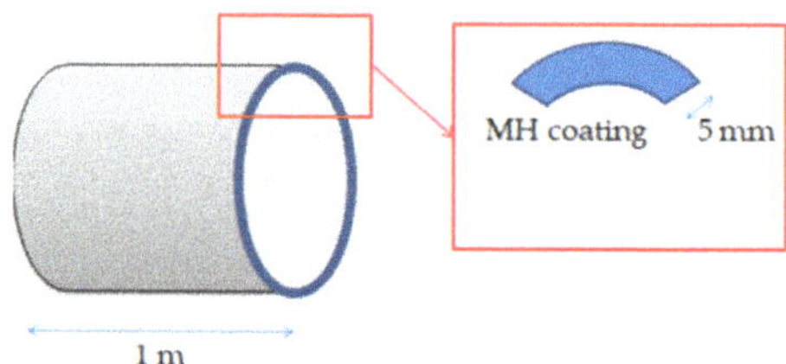

Figure 1. Schematic representation of the magnesium hydroxide coating (MHC) in a sewer pipe.

2.2. Cost-Comparative Analysis

The entire initial analysis of the methodology and the parameters to be studied are presented in detail by [19,33]. Consequently, this paper's insight includes the least cost analysis framework assessment for the sewage system and all the relevant actions included in the maintenance and repair stage for controlling corrosion. The current framework is multilateral and contains a step by step evaluation regarding economic and social extensions, while providing rational conclusions. Hence, the system boundaries will be depicted together with the goals of the approach similar to the LCA perspective. Next in order will be the inventory analysis, and in continuation the formulas for the economic assessment will be illustrated, in order to proceed to the results section.

2.2.1. Goal and Scope Definition

The main aim of this paper is to investigate the costs of maintenance via two MIC mitigation strategies and to provide recommendations for each case, based on a least cost analysis methodological framework. The system boundaries were set to a fifty-year continuous procedure of maintenance via a net present value (NPV) formula for every mitigation strategy, while PVC installation is also considered as a third option. The economic system is included in the social and natural system. In the latter, it is not possible to record the external influences that come in and out of the system, as it is most likely that they cannot be easily quantified (e.g., disturbance cost which is a social impact). Therefore, the system is divided into two parts, in which the primary section represents the internal costs, as recognized by [34] and the secondary section illustrates external costs in the form of delay due to the intervention of maintenance [33] (Figure 2).

Regarding the functional unit of the relevant system, it is usually set to flow rate of sewage (effluent), though the length or the weight of sewer pipeline is a usually implemented functional unit for similar studies [15]. Due to insufficient data for the flow rate for every town, 1 km of implemented maintenance actions is utilized in this study. Furthermore, the current format is used because the assessment is strictly targeted to the actions of maintenance, and not to the whole project design and operation stage of the sewerage network. Finally, the quantity of inputs differs based on the pipe's size, as well as the amount of labor hours and machinery depreciation. Given that, an inventory is formulated with all the relevant inputs and assumptions of this research.

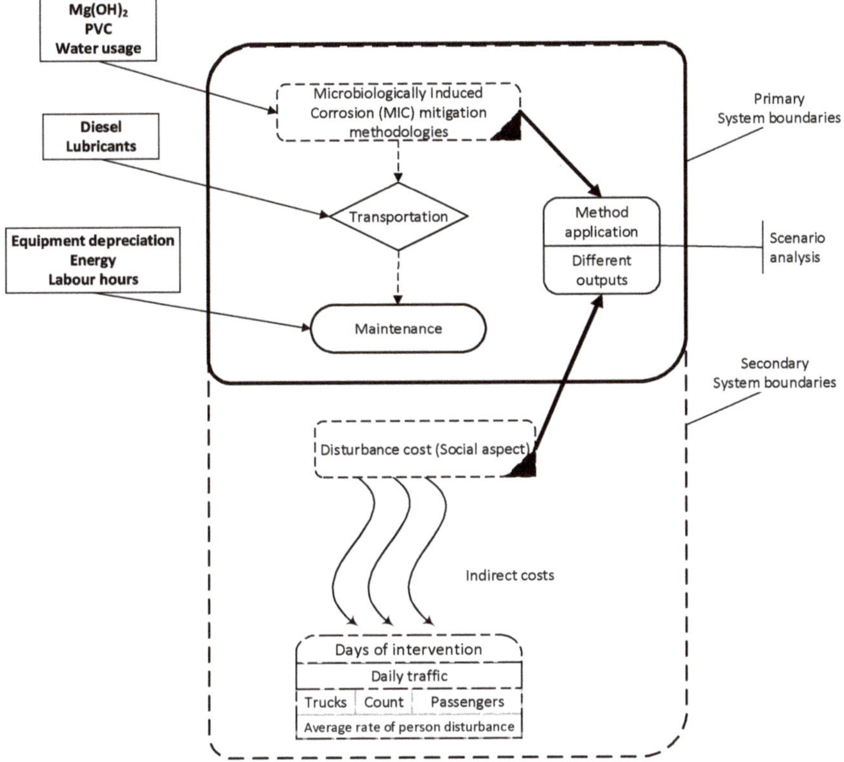

Figure 2. System Boundaries.

2.2.2. Inventory Analysis

The formation of an inventory is a complicated procedure which demands analytical depiction of the designated goals and data. In comparison to an LCA inventory, the tangible aspect of economic datasheets develops an easily understandable environment for our approach, from which the final results can be interpreted without difficulty. This signifies the importance of least cost analysis in a system of controlling corrosion methods, which is mostly based on activities (maintenance and repair), and to a lesser degree on the production of solid materials (e.g., the construction phase).

More specifically, costs are divided into different driving forces (e.g., hours of work, number of employees, amount of consumables, etc.) and then classified into different cost categories (e.g., EUR/labor hour, EUR/mechanical labor, EUR/worksheet, etc.). The inventory is enriched with data from field measurements [7], whilst specific coefficients are taken from official records (e.g., euros per hour of labor). Furthermore, valid assumptions—based on the primary data—should be made, in order to have credible results.

In conjunction with the abovementioned, an inventory analysis is depicted in Table 1. Several assumptions have been made in relation to the radius of the maintenance interventions. Specifically, in Thessaloniki the radius was set to 50 km, in Serres to 30 km, and in Kozani and Florina to 20 km, based on the region's size. The economic life of the maintenance intervention was set to 50 years, since the service life of PVC as sewer material has a maximum of 50 years [33]. The prices for PVC installations were taken from a national restricted call for tenders in the Greek region. Finally, all the other values have been obtained from specific references and/or current prices from the Greek territory.

Table 1. Inventory analysis.

Assumptions	Unit	Values	Remarks
Area radius	km	20–50	Different scenarios based on city size
Expected economic life	yr	50	[19]
Inputs			
Labour	€/h	6.33	National payscale
Water consumption	€/m^3	0.95–1.11	Based on the selling price of each MWSE
Mg(OH)$_2$ Slurry	€/m	2.65–10.60	Based on the sewer diameter
Diesel	€/l	1.5	
Lubricants	€/l	3.275	7% of diesel
Energy	€/kWh	0.13001	National electricity price
Outputs			
NPV for each corrosion mitigation strategy	€/km		Figures 3–5
PVC replacement	€/km	361,567–569,320	Based on the Concrete Pipe Size
Equipment depreciation			
Sprayer	€/h	1.26	ABC software http://www.abc.aua.gr/
Lorry 2 T	€/h	10.86	ABC software http://www.abc.aua.gr/
Disturbance (Indirect impacts)			
Annual daily traffic	Vehicles/day	236–4128	[35]
Time delay	h	0.15–0.5	Based on the controlling method
Days of each project	days	-	Based on the duration
Person labor hour	€/h	6.33	National payscale
Freight delay	€/h	17.97	National payscale
Vehicle passenger traffic	%	97	[33]
Vehicle freight traffic	%	3	[33]
Occupancy of vehicle	Persons	1.2	[33]

2.2.3. Economic Assessment Formulas

Several costs in relation to acquisition, energy consumption, maintenance, operation, end of life costs, and various other costs over the life cycle of a product and/or service should be considered. In this context, this methodological framework accounting for all these relevant costs of a product or an activity is discounted over its life existence. Accounting the costs for a sewer system is based on a simple equation for each activity *i*, aggregating different amounts of costs (CC: Cumulative Costs) [21,22,36]:

$$CC = C_{ini} + C_{op} + C_m + C_{en} + C_{ind} - C_{rv} \qquad (4)$$

where C_{ini} represents the initial costs of commissioning, purchasing, and installing inputs. C_{op} depicts the operation costs, C_m the maintenance costs, C_{en} the energy costs, and C_{rv} the residual value of the investment. Regarding C_{ind}, there are several indirect costs included in a construction project such as environmental, social, downtime, and decommissioning costs. In order to untangle the cost of a project over its lifetime, specific future costs should be depicted in equivalent values of today. Furthermore, replacement costs are represented as onetime actions, while others like maintenance costs are more frequent. This is where the present value (PV) is introduced, based on equivalent costs of actions at a current or present time. Consequently, this would be the amount of money that should be set aside today, in order to face future costs of the desired design project. The PV calculation is illustrated below:

$$PV = \frac{FV}{(1+r)^n} \qquad (5)$$

where *FV* is the future value, *r* is the rate of return, and *n* is the number of periods. When compounding the inflation/interest rate, a general formula is illustrated for drainage projects by the Concrete Pipe Association of Australasia (ACPA) [33] as presented below:

$$FC = PC\,(1+I)^n \tag{6}$$

or

$$PC = FC\,(1+i)^n \tag{7}$$

where FC is the future cost, PC is the present cost, i is the interest rate, I is the inflation rate, and n is the number of periods. Usual maintenance actions for concrete sewers include the removal of debris, silt removal, flushing, and the repair of damage. For this research focus is given to the maintenance actions in relation to the MIC in concrete sewers. This includes two different maintenance actions, namely flushing with high-pressure water, and spraying of MHS. The maintenance cost equation is calculated below, according to ACPA [33], based on the durability variation of the applicable products:

$$MC = A_{mc}\frac{1-F^n}{\frac{1}{F}-1} \tag{8}$$

where MC represents the maintenance cost, A_{mc} is the annual maintenance cost and n is the years of service life. The value F depicts the interaction between the inflation and the discount rate as illustrated below:

$$F = \left(\frac{1+I}{1+i}\right)^n \tag{9}$$

Apart from the direct economic impacts due to sewer maintenance, there are also indirect impacts as mentioned before. The economic impact due to disturbances during road closures is calculated as a secondary part of this research, since it could be a significant cost to the social life of an area. Within this framework the disturbance formula is integrated and is presented below [33]:

$$Disturbance\ cost = AADT * t_h * d_p * \left(c_{vh} * v_v * v_{of} + c_{fh} * v_f\right) \tag{10}$$

where
$AADT$ = annual average daily traffic of the relevant roadway
t_h = average increase in delay in hours for every vehicle per day
d_p = number of days the project will take
c_{vh} = average rate of person-labor hour, in euros per hour
v_v = percentage of passenger vehicles traffic
v_{of} = vehicle occupancy factor
c_{fh} = average rate of freight-delay, in euros per hour
v_f = percentage of truck traffic

3. Results

The cities of the case studies are presented in Table 2. More specifically, Thessaloniki is the second largest metropolis in Greece, which plays a significant role in the Balkan region as a trade center. Serres and Kozani share the same statistics in population and Florina is the smallest of all with many disadvantaged areas due to the high altitude. This creates an uneven situation among them regarding the deterioration of their sewer system, inducing manifold problems due to different needs. In Table 2, the "Peak Population Equivalent" data was taken by the monitoring database of the special secretariat for water, Ministry of Environment and Energy [37]. In addition, in Table 2, a combined system is a system which carries both surface run-off and wastewater, while a separate system carries the municipal wastewater and surface run-off separately. The concrete pipe size of Table 2 is one of the main sewer diameters of each city.

Table 2. Details of case study cities.

Town	Town Number on Map	Peak Population Equivalent	Type of System	Concrete Pipe Size
Thessaloniki	1	900,000	Combined	1000 mm
Serres	2	79,000	Combined	800 mm
Kozani	3	46,000	Combined	250 mm
Florina	4	20,000	Separate	400 mm

In Figure 3, the accumulated costs for all cities for 50 years of corrosion prevention strategies are presented. The current inflation rate of Greece is integrated, which is 0.9%, while for the discount rate the European recommendation (Commission Recommendation (EU) 2019/1659) for a social discount rate in economic analysis (3%) is followed. The major difference in Figure 3 is the depiction of costs in a chronological order. Although the maintenance costs of MHS and water flushing are presented in PV form through time, the cost of PVC installation is depicted as a one-time investment cost. This classification has been made due to the permanent character of PVC installations in comparison to mitigation methodologies used in the maintenance of concrete sewers, which should take place many times in the years to come. In this manner, the significance of the decision-making regarding the right option between biocorrosion mitigation techniques could be illustrated. If the selected option is PVC, then the right time to proceed with the current project is as soon as possible, since the costs will increase drastically.

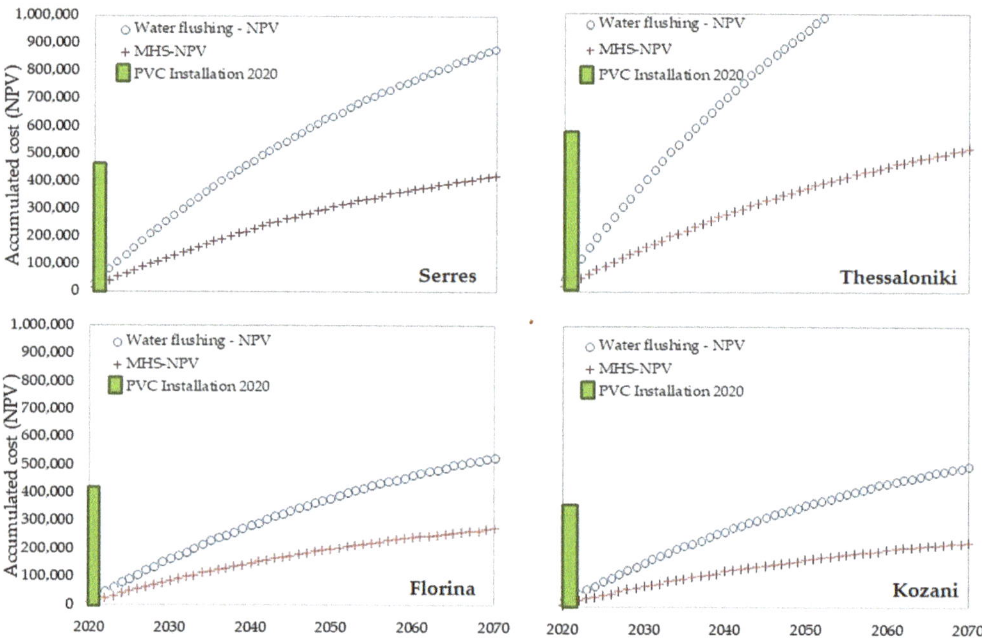

Figure 3. Accumulated costs for 50 years' corrosion prevention strategies.

On the other hand, for a short term strategy, the application of MHS and even water flushing (for a very short period) is a much more attractive investment for the MWSEs. In Kozani, the accumulated costs for a 50-year corrosion prevention strategy plan for concrete pipes with diameter 250 mm are presented. It can be observed that MHS spraying is the best option. The value of water flushing, although much more costly compared to MHS, is less expensive compared to the new PVC installation in Kozani, a middle-sized town of 150,000 residents which is experiencing concrete biocorrosion problems [7], and where the local MWSE is using high-pressure water flushing as a biocorrosion mitigation technique. In the study case of Florina, the results showed similar trends to those of Kozani.

However, in the cases of Serres and Thessaloniki, the high-pressure water flushing is the worst option, while MHS spraying seems the most sustainable plan. After 25 years in Serres and after 20 years in Thessaloniki, the net present value of high-pressure water flushing is shown to reach the PVC installation present costs. These results can be explained by the following reasons: (i) the amount of maintenance labor hours in Thessaloniki is much higher compared to the cities, (ii) Thessaloniki has the largest pipe diameters among the cities and that makes water-flushing the most expensive method of all.

In Figures 4 and 5 the NPV graphs for 50 years' use of high-pressure water flushing and for the application of MHC onto sewer pipes are presented accordingly, with a small alteration to the inflation rates. Greece faced severe economic uncertainty from 2008 onwards, with financial indicators constantly changing. Deflation has had a significant impact on the Greek economy, with continuously falling prices and decreasing consumption. While the country still tries to overcome economic obstacles, the inflation rate (IR) is noticeably lower than the average inflation rate in the EU. Therefore, the inflation rate will increase in the years to come, since the simulations follow this trend, and the authors have taken this into consideration. IRs of 0.9% and 1.6% are plotted, as 1.6% is considered as an EU average. When it comes to the use of high-pressure water, it can be observed that in the case of IR 1.6%, installing PVC sewer pipes is currently a more cost-efficient decision than using high-pressure water for 50 years.

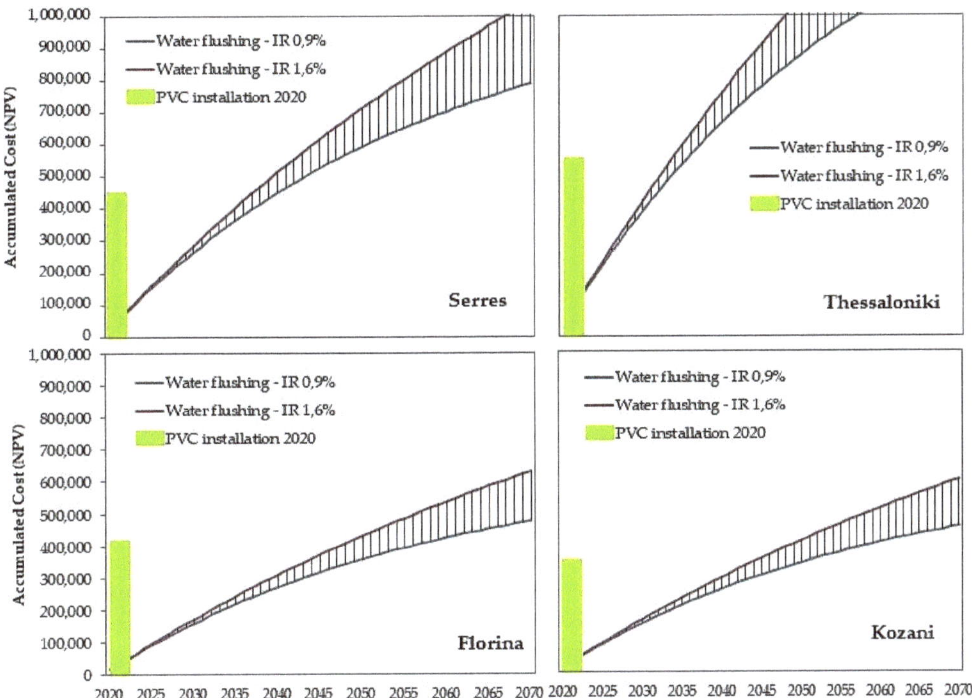

Figure 4. Net present value (NPV) graphs for high-pressure water flushing for 50 years for inflation rates (IR) 0.9% and 1.6%.

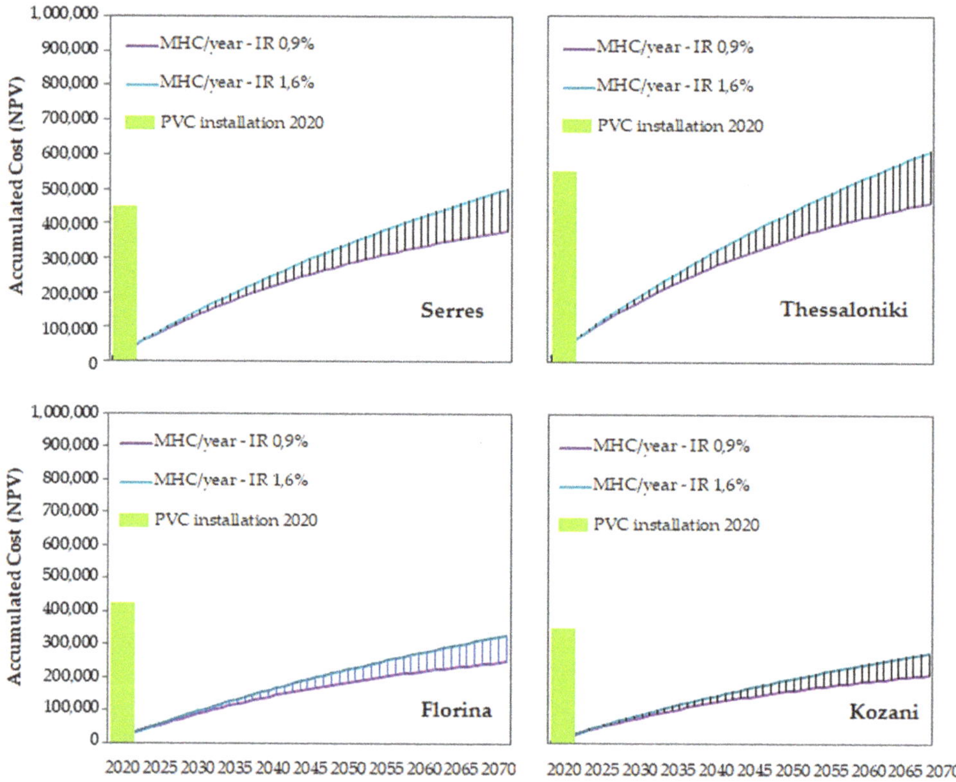

Figure 5. NPV graphs for MHC/year for 50 years for IR 0.9% and 1.6%.

In the case of MHS, Serres and Thessaloniki showed that a possible PVC installation at the current time might be a reasonable option. In the case of Florina and Kozani, even with the case of a 1.6% IR, the use of MHS is a cost-efficient option. It should be stated that the indirect cost of disturbances experienced by the residents of the cities (including lost time and road usage) is also vital in evaluating the total installation cost of PVC pipes. That way, the PVC pipe installation cost could increase to EUR 1 million, making the PVC installation a much less favorable option.

It is important to note that a combination of methods could be a significant advantage for decision makers, selecting among short-term and long-term solutions. More specifically, the choice of MHS for five years and in continuation with the installation of PVC for a long-term solution could be advantageous in case of a temporary low level of working capital. It could be an asset for local authorities in order to estimate and distribute future monetary plans to this direction, especially for smaller cities, as shown in Figure 5 even with higher values of IR.

Regarding the secondary part of the indirect social impacts on the community, in Table 3 these externalities are presented based on different controlling methods and the area of application. The non-stop nature of water flushing with maintenance interventions every month has a significant social cost in every area, though in Thessaloniki the economic damage could exceed the other values by a large margin. The second least economic option is the PVC installation, in which although the maintenance should disturb people only once throughout a 50-year period, this intervention would also close entire roadways for months. Finally, the application of MHS has the least economic impact on the local economy. For the application of MHS once per year, the economic cost is the lowest for every city, with losses between EUR 45,906 and EUR 802,969.

Table 3. Indirect costs in euro from disturbance in a 50-year maintenance project based on different controlling methods.

Controlling Method	Kozani	Thessaloniki	Serres	Florina
Water flushing	2,121,665 €	15,667,682 €	2,379,757 €	895,730 €
MHS/year	108,735 €	802,969 €	121,963 €	45,906 €
PVC installation	265,208 €	1,468,845 €	223,102 €	83,975 €

4. Limitations of This Study

The current approach targets only specific economic aspects and does not cover the whole procedure. This paper should be useful for a first cost-comparison estimation of the biocorrosion mitigation methodologies, and at a later stage a detailed life cycle costing or a techno-economical study should follow. Technical aspects regarding the "performance" of the biocorrosion mitigation measures were not studied. Material changes are to be expected over a service life of decades, especially if one of the two preventive measures is regularly applied to concrete surfaces. The durability of the material and also the long-term effectiveness should be considered. All these aspects should be considered when a local wastewater treatment facility agrees on a corrosion mitigation strategy.

5. Conclusions

In order to reach a sustainable maintenance treatment strategy in the sewer systems, a cost-comparison analysis was performed to examine biocorrosion mitigation methodologies. Spraying with MHS was considered to be the most cost-effective strategy to mitigate MIC. The main issue for local decision-makers is the time of decision-making. Today, the application of MHS is the most advantageous maintenance technique, based on direct and indirect economic assessment. Nevertheless, if the local authorities do not take into consideration the indirect social impacts, PVC installation is considered to be a cost-effective solution for larger cities. Furthermore, the least cost-efficient approach to the current biocorrosion problem in concrete sewers is water flushing, due to both direct and indirect economic parameters.

As future work, the authors will focus on the development of a universal financial assessment tool for the investigation of economic impacts related to corrosion controlling methods, integrating relevant actions of a circular economy model framework proposition [38]. Furthermore, future studies could focus on the installment of new liners inside old concrete pipes instead of the aforementioned biocorrosion mitigation strategies, since this could be considered to be a sustainable option in a circular economy as well.

Author Contributions: Conceptualization, G.F. and E.T.; methodology, G.F., E.T. and P.S.; software, E.T.; validation, G.F. and E.T.; formal analysis, G.F. and E.P.; investigation, G.F., E.T. and E.P.; resources, P.S.; data curation, G.F., E.T.; writing—original draft preparation, G.F. and E.T.; writing—review and editing, E.P. and P.S.; visualization, G.F. and E.T.; supervision, P.S.; project administration, P.S.; funding acquisition, P.S. All authors have read and agreed to the published version of the manuscript.

Funding: This research has been co-financed by the European Union and Greek national funds through the Operational Program Competitiveness, Entrepreneurship and Innovation, under the call RESEARCH—CREATE—INNOVATE (project code: T1EDK-02355-title 'Novel Coating Materials for Corrosion Protection of Sewer Network Pipes').

Conflicts of Interest: The authors declare no conflict of interest.

References

1. Islander, R.L.; Devinny, J.S.; Mansfeld, F.; Postyn, A.; Shih, H. Microbial Ecology of Crown Corrosion in Sewers. *J. Environ. Eng.* **1991**, *117*, 751–770. [CrossRef]
2. O'Dea, V. Understanding biogenic sulfide corrosion. *Mater. Perform.* **2007**, *46*, 11.
3. Wang, T.; Wu, K.; Kan, L.; Wu, M. Current understanding on microbiologically induced corrosion of concrete in sewer structures: A review of the evaluation methods and mitigation measures. *Constr. Build. Mater.* **2020**, *247*, 118539. [CrossRef]

4. Grengg, C.; Mittermayr, F.; Ukrainczyk, N.; Koraimann, G.; Kienesberger, S.; Dietzel, M. Advances in concrete materials for sewer systems affected by microbial induced concrete corrosion: A review. *Water Res.* **2018**, *134*, 341–352. [CrossRef] [PubMed]

5. Zhang, L.; De Schryver, P.; De Gusseme, B.; De Muynck, W.; Boon, N.; Verstraete, W. Chemical and biological technologies for hydrogen sulfide emission control in sewer systems: A review. *Water Res.* **2008**, *42*, 1–12. [CrossRef] [PubMed]

6. Roghanian, N.; Banthia, N. Development of a sustainable coating and repair material to prevent bio-corrosion in concrete sewer and waste-water pipes. *Cem. Concr. Compos.* **2019**, *100*, 99–107. [CrossRef]

7. Fytianos, G.; Baltikas, V.; Loukovitis, D.; Banti, D.; Sfikas, A.; Papastergiadis, E.; Samaras, P. Biocorrosion of Concrete Sewers in Greece: Current Practices and Challenges. *Sustainability* **2020**, *12*, 2638. [CrossRef]

8. Karageorgos, P.; Latos, M.; Kotsifaki, C.; Lazaridis, M.; Kalogerakis, N. Treatment of unpleasant odors in municipal wastewater treatment plants. *Water Sci. Technol.* **2010**, *61*, 2635–2644. [CrossRef]

9. Bravo, L.; Ferrer, I. Life Cycle Assessment of an intensive sewage treatment plant in Barcelona (Spain) with focus on energy aspects. *Water Sci. Technol.* **2011**, *64*, 440–447. [CrossRef]

10. Gaterell, M.; Griffin, P.; Lester, J. Evaluation of Environmental Burdens Associated with Sewage Treatment Processes Using Life Cycle Assessment Techniques. *Environ. Technol.* **2005**, *26*, 231–250. [CrossRef]

11. Gallego, A.; Hospido, A.; Moreira, M.T.; Feijoo, G. Environmental performance of wastewater treatment plants for small populations. *Resour. Conserv. Recycl.* **2008**, *52*, 931–940. [CrossRef]

12. Hospido, A.; Moreira, M.T.; Feijoo, G. A comparison of municipal wastewater treatment plants for big centres of population in Galicia (Spain). *Int. J. Life Cycle Assess.* **2007**, *13*, 57. [CrossRef]

13. Ortiz, M.; Raluy, R.G.; Serra, L.M. Life cycle assessment of water treatment technologies: Wastewater and water-reuse in a small town. *Desalination* **2007**, *204*, 121–131. [CrossRef]

14. Sustainable Solutions Corporation, Life Cycle Assessment of PVC Water and Sewer Pipe and Comparative Sustainability Analysis of Pipe Materials Royersford, USA. 2017. Available online: https://www.uni-bell.org/files/Reports/Life_Cycle_Assessment_of_PVC_Water_and_Sewer_Pipe_and_Comparative_Sustainability_Analysis_of_Pipe_Materials.pdf (accessed on 16 March 2020).

15. Kim, D.; Yi, S.; Lee, W. Life cycle assessment of sewer system: Comparison of pipe materials. In Proceedings of the 2012 World Congress on Advances in Civil, Environmental, and Materials Research (ACEN' 12), Seoul, Korea, 26–30 August 2012.

16. Gjorv, O.E.; Sakai, K. *Concrete Technology for a Sustainable Development in the 21st Century*; CRC Press: Boca Raton, FL, USA, 2003; pp. 332–340.

17. Æsøy, A.; Østerhus, S.; Bentzen, G. Controlled treatment with nitrate in sewers to prevent concrete corrosion. *Water Supply* **2002**, *2*, 137–144. [CrossRef]

18. Saleh, M.Y.; EL-Zahar, M.M.H.; Ashour, N.M.M. Application of Life Cycle Cost Analysis Method on Wastewater Treatment Plants In Egypt. In Proceedings of the Twenty-first International Water Technology Conference, IWTC21, Ismailia, Egypt, 28–30 June 2018.

19. Stein, D. *Rehabilitation and Maintenance of Drains and Sewers*; Ernst & Sohn Wiley: Berlin, Germany, 2001; 804p.

20. UN-Habitat. *State of the World's Cities 2008/2009–Harmonious Cities*; UN-Habitat: Nairobi, Kenya, 2008.

21. XYLEM. WHITE PAPER Life Cycle Costs for Wastewater Pumping Systems June 2015. Available online: https://www.xylem.com/siteassets/support/tekniska-rapporter/white-papers-pdf/life-cycle-costs-lcc-for-wastewater-pumping-systems.pdf (accessed on 17 March 2020).

22. Concrete Pipe Associationof Australasia. Life Cycle Cost Analysis in Drainage Projects. *Technical Brief.* 2015. Available online: https://www.cpaa.asn.au/images/publications/technical_briefs/life%20cycle%20drainage.pdf (accessed on 17 March 2020).

23. Yoshida, H.; Christensen, T.H.; Scheutz, C. Life cycle assessment of sewage sludge management: A review. *Waste Manag. Res.* **2013**, *31*, 1083–1101. [CrossRef]

24. Abraham, D.M. Life Cycle Cost Integration for the Rehabilitation of Wastewater Infrastructure. *Constr. Res. Congr.* **2003**, 1–9. [CrossRef]

25. Årskog, V.; Fossdal, S. Life-Cycle Assessment of Repair and Maintenance Systems For Concrete Structures. In Proceedings of the International Workshop on Sustainable Development and Concrete Technology, Beijing, China, 20–21 May 2014.

26. Tsagarakis, K.P.; Mara, D.; Angelakis, A.N. Application of Cost Criteria for Selection of Municipal Wastewater Treatment Systems. *Water Air Soil Pollut.* **2003**, *142*, 187–210. [CrossRef]

27. Wu, L.; Hu, C.; Liu, W.V. The Sustainability of Concrete in Sewer Tunnel—A Narrative Review of Acid Corrosion in the City of Edmonton, Canada. *Sustainability* **2018**, *10*, 517. [CrossRef]

28. Noeiaghaei, T.; Mukherjee, A.; Mistri, A.; Chae, S.-R. Biogenic deterioration of concrete and its mitigation technologies. *Constr. Build. Mater.* **2017**, *149*, 575–586. [CrossRef]

29. Sydney, R.; Esfandi, E.; Surapaneni, S. Control concrete sewer corrosion via the crown spray process. *Water Environ. Res.* **1996**, *68*, 338–347. [CrossRef]

30. Nielsen, A.H.; Vollertsen, J.; Jensen, H.S.; Wium-Andersen, T.; Hvitved-Jacobsen, T. Influence of pipe material and surfaces on sulfide related odor and corrosion in sewers. *Water Res.* **2008**, *42*, 4206–4214. [CrossRef] [PubMed]

31. Sun, X.; Jiang, G.; Chiu, T.H.; Zhou, M.; Keller, J.; Bond, P.L. Effects of surface washing on the mitigation of concrete corrosion under sewer conditions. *Cem. Concr. Compos.* **2016**, *68*, 88–95. [CrossRef]

32. Crownmag. Innovative Coating Materials for Corrosion Protection of Sewer Network Pipes. Available online: http://crownmag.gr/en/ (accessed on 23 March 2020).

33. American Concrete Pipe Assocation. Least Cost Analysis. Resource 07-130 (06/07) 5M 2007. Available online: https://www.concretepipe.org/wp-content/uploads/2015/05/LCA-Brochure.pdf (accessed on 23 March 2020).

34. Rebitzer, G.; Hunkeler, D. Life cycle costing in LCM: Ambitions, opportunities, and limitations. *Int. J. Life Cycle Assess.* **2003**, *8*, 253–256. [CrossRef]

35. Mitsakis, E.; Stamos, I.; Grau, J.M.S.; Chrysochoou, E.; Iordanopoulos, P.; Aifadopoulou, G. Urban Mobility Indicators for Thessaloniki. *J. Traffic Logist. Eng.* **2013**, *1*, 148–152. [CrossRef]

36. Lemer, A. Life Cycle Costing: For the Analysis, Management and Maintenance of Civil Engineering Infrastructure. *Constr. Manag. Econ.* **2015**, *33*, 689–691. [CrossRef]

37. Wastewater Treatment Plants. Monitoring Database. Special Secretariat for Water. Ministry of Environment and Energy. Available online: http://astikalimata.ypeka.gr/Services/Pages/WtpViewApp.aspx (accessed on 24 March 2020).

38. Smol, M.; Adam, C.; Preisner, M. Circular economy model framework in the European water and wastewater sector. *J. Mater. Cycles Waste Manag.* **2020**, *22*, 682–697. [CrossRef]

sustainability

Article

Lifecycle Design of Disruptive SCADA Systems for Waste-Water Treatment Installations

Stelian Brad [1,2,*], **Mircea Murar** [1], **Grigore Vlad** [3], **Emilia Brad** [1] and **Mariuța Popanton** [1,3]

[1] Department of Engineering Design and Robotics, Technical University of Cluj-Napoca, Bvd. Muncii 103–105, 400641 Cluj-Napoca, Romania; mircea.murar@muri.utcluj.ro (M.M.); emilia.brad@muri.utcluj.ro (E.B.); maria.popanton@icpebn.ro (M.P.)

[2] Cluj IT Cluster, Memorandumului 29, 400441 Cluj-Napoca, Romania

[3] Research and Development Institute for Environmental Protection Technologies and Equipment, Parcului 7, 420035 Bistrița, Romania; vlad@icpebn.ro

* Correspondence: stelian.brad@staff.utcluj.ro; Tel.: +40-730-017-126

Abstract: Capacity to remotely monitor and control systems for waste-water treatment and to provide real time and trustworthy data of system's behavior to various stakeholders is of high relevance. SCADA systems are used to undertake this job. SCADA solutions are usually conceptualized and designed with a major focus on technological integrability and functionality. Very little contributions are brought to optimize these systems with respect to a mix of target functions, especially considering a lifecycle perspective. In this paper, we propose a structured methodology for optimizing SCADA systems from a lifecycle perspective for the specific case of waste-water treatment units. The methodology embeds techniques for handling entropy in the design process and to assist engineers in designing effective solutions in a space with multiple constrains and conflicts. Evolutionary multiple optimization algorithms are used to handle this challenge. After the foundation of the theoretical model calibrated for the specific case of waste-water treatment units, a practical example illustrates its applicability. It is shown how the model can lead to a disruptive solution, which integrates cloud computing, IoT, and data analytics in the SCADA system, with some competitive advantages in terms of flexibility, cost effectiveness, and increased value added for both integrators and beneficiaries.

Keywords: waste-water management; SCADA; design optimization; remote control; IoT; cloud computing; disruptive innovation; lifecycle

Citation: Brad, S.; Murar, M.; Vlad, G.; Brad, E.; Popanton, M. Lifecycle Design of Disruptive SCADA Systems for Waste-Water Treatment Installations. *Sustainability* **2021**, *13*, 4950. https://doi.org/10.3390/su13094950

Academic Editor: Charikleia Prochaska

Received: 6 April 2021
Accepted: 26 April 2021
Published: 28 April 2021

Publisher's Note: MDPI stays neutral with regard to jurisdictional claims in published maps and institutional affiliations.

1. Introduction

Waste-water treatment (WWT) plays a tremendous role for environmental health and quality of life. WWT installations are organized into geographically distributed technological units or complete WWT plants that are connected by means of communication technologies to a regional or central control center to deliver a holistic approach of real-time control and monitoring actions upon the special technological processes [1].

Besides the necessary technological innovations that are focused on cost-effective and quality-effective purification of waste-water before releasing in rivers or directing to secondary use (e.g., in agriculture or other industrial sectors), continuous monitoring and control of treatment installations to keep within nominal working parameters, as well as to act preventively or proactively against failures, is of the same relevance.

Supervisory Control and Data Acquisition (SCADA) are specialized control system architectures dedicated for performing this job [2]. They comprise computers, networks for data communication, Programmable Logic Controllers (PLCs) that interface sensors and actuators with process supervisory management units, and other control units to interface processes with installations and equipment [3,4]. The key feature of a SCADA system is the capacity to perform a supervisory operation over a variety of other proprietary devices, using standard communication protocols [4]. SCADA systems embed high-level human-process interfaces and are characterized by a wide range of monitoring and controlling functions, including trend view of parameters, reports, alarms, and notifications [3–5]. SCADA systems can be designed both for small and very large installations [6,7]. Figure 1 introduces the generic SCADA architecture [2,3,7].

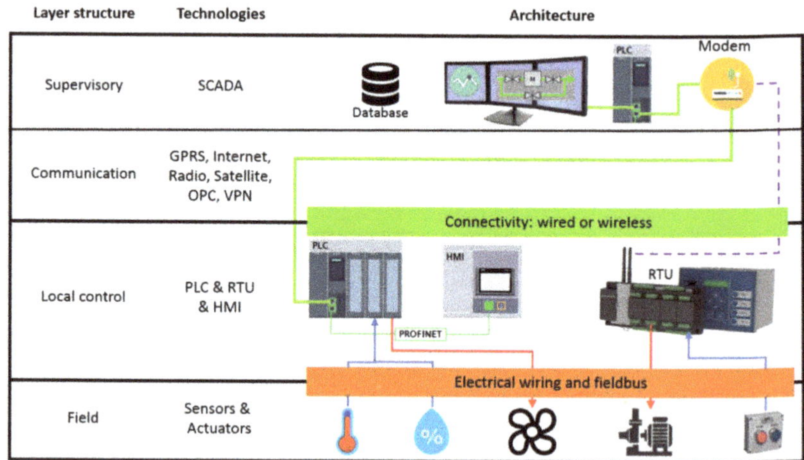

Figure 1. The generic SCADA architecture.

According to Figure 1, field equipment (e.g., sensors, actuators, pumps, switches, turbines, etc.) is connected to PLCs or Remote Terminal Units (RTUs) in a local network, where input data are gathered and processed. Actions are triggered based on the result of a control logic which runs inside PLCs or RTUs [2,5,8]. Critical information is sent in real-time from geographically distributed units to a remote monitoring unit via proprietary communication protocols on top of communication technologies provided by communication companies [5]. Data are preprocessed by a master control unit and prepared for visualization on Human-Process Interfaces (HPI) and saved in a database from the control center. An operator analyzes data and can trigger remote actions to terminal units when this is necessary [9].

2. Background

This section covers the research focus of the current paper. In the first part, the main building blocks for engineering a SCADA system are introduced. Afterwards, drawbacks of these systems are underlined. The section ends with the foundation of the research scope and frames the area for analyzing the state of the art in the field.

From a conceptual perspective, we must see a SCADA system as a bridge between at least one remote operation unit (e.g., sensor, actuator) located in the field and a data visualization and management unit. Through the SCADA system, automation logic is realized, and data are exchanged between the field and the supervisory unit [9,10]. Therefore, the major engineering blocks of a SCADA system are data generation and management, connectivity, and commissioning [11]. Thus, the control center includes design and development of SCADA application and human-process interface (HPI) to visualize processed data that are received from remote automation systems. Connectivity between SCADA application and remote automation systems involves setup and configuration of the communication hardware and services, identification and hiring services from communication providers that can ensure the network coverage, static IPs, and VPN gates for secured communication. Commissioning puts the whole system into operation, tests and validates data and functionalities, etc. This job requires a great number of skilled human resources and represents one of the major challenges in SCADA system engineering.

Considering the plethora of components, modules, and technologies (e.g., communication) required to setup a SCADA system, different parts of a SCADA architecture are usually developed and engineered by different vendors or subcontractors. This process comes up with a series of drawbacks, such as [7,12,13]:

- Lack of homogeneity to the level of SCADA applications and automation solutions are a very usual situation in the case of WWT plants, with challenges in terms of intervention, maintenance, and servicing, as well as in terms of integration and data interoperability.
- High level of heterogeneity for data representation, variable naming, data processing, equipment control, signaling, and visualization, with challenges and subsequent complications in terms of data modelling for analytics, aggregated preventive maintenance, and agility in scaling-up monitoring and control of WWT plants.
- Difficulties to receive support or even missing assistance in the case of software, hardware, or operating systems that become technologically obsolete, but still functional in the WWT plant, with subsequent blockages for renewing or updating SCADA system applications that run on the top.
- High costs for engineering, commissioning, and lifecycle operation of SCADA applications, because of the system's heterogeneous character and necessity to collaborate with more integrators.
- Complications for upscaling and updating the system, even for executing small adjustments that frequently occur, because more vendors or integrators must be simultaneously involved.

Stricter compliance measures are imposed by regulatory bodies on the water industry to increase the reliance of data delivered from the treatment processes. This requires different approaches in the development of SCADA systems and automation units. To increase the quality of life and safety of citizens, WWT installations must be distributed in as many places as possible within the waste-water network of a city. Instead of having a big plant, the smarter, more resilient, and safer way is to have a grid of remote controllable WWT units. This also facilities agility in operation, control of waste-water sources, as well as flexibility in operation, including updating and upscaling. For a smart city, this generic architecture is desirable [14,15]. However, integrating current solutions of SCADA systems in a smart city network of WWT would be very labor intensive, time consuming, and costly [16,17]. Moreover, due to the large variety of technologies that might be implemented, as well as due to various standards and various control strategies, data management and processing in a grid of WWT installations might be also technically difficult [1,18].

Due to advancements in computing and remote communication technologies, such as cloud computing, big data management and IoT, consideration of these new technologies as part of a grid of WWT installations could provide several advantages, including better serviceability, higher resilience, and better lifecycle orientation [19,20], as well as considerable mitigation of current drawbacks of SCADA systems [21,22].

Therefore, the goal of this paper is to introduce a methodology that systematically manages a lifecycle oriented, multi-criteria design process of a SCADA system for a grid of smart WWT installations that embeds cloud computing and data analytics [20,22] instead of classical remote operating systems, to facilitate the inclusion of artificial intelligence models for preventive maintenance and decision making over the system's lifecycle. In terms of the system's optimal design, a set of criteria that covers all lifecycle phases are considered: efficiency in operation, enhanced options for data matching, redundancy for higher safety standards, longer life, initial cost reduction with engineering and commissioning, as well as cost reduction over lifecycle operation.

Hence, the remainder of the paper is organized as follows. In Section 3, related work on SCADA system architectures for WWT installations is investigated using databases of publications and combinations of keywords for searching and filtering. Results are used to formulate the gap and to indicate the novelty brought by this research. In Section 4, the research methodology is presented into details. Because multiple optimization criteria are involved and because some of them are in conflict each other, an evolutionary design algorithm that integrates systematic problem-solving techniques is proposed. It looks for the effective path for investigating architectural construction and technology selection such as to best satisfy each optimization criterion. Because the algorithm deals with system optimization—not parameter optimization—at this stage of the design we must operate with qualitative optimization, not with a quantitative one. Section 5 illustrates the application of the methodology to design a disruptive SCADA system. Disruption must be understood according to its definition, in the sense of defining a solution that can replace traditional SCADA systems in a smart grid of WWT installations, at least for the category of low-end beneficiaries, such as villages, small districts, isolated resorts, small towns, as well as local WWT installations of industrial facilities, but also with capacity to be integrated in large WWT plants. The paper ends with a section of discussions around the results of the case study and a section of conclusions that indicates advantages and limitations of the proposed solution, a comparative analysis with traditional SCADA systems, findings from this research and highlights on future researches in relation to this topic.

3. Related Work

To analyze the current researches in cloud computing-based SCADA systems we have investigated papers published on the Web of Sciences, IEEE Xplore, and SCOPUS databases. Searching the Web of Science database with the combination (SCADA" and "IoT" and "cloud computing"), only 11 references were returned. From this set, 6 papers focus on cybersecurity issues, which are out of the scope of this research. In the remaining 5 papers, only 2 deal with water installations, and one with aspects on SCADA initial steps of design. Thus, these 3 papers were found to be relevant for further analysis in relation with the focus of this paper. With the same combination of keywords, IEEE Xplore returned 19 references. Eliminating those that are also indexed in the Web of Science, none of the remaining 8 papers fit into the scope of this research. The SCOPUS database returned 20 titles for the same combination of keywords, but the relevant references are the same as indexed in the Web of Science. The conclusion is that only few researches have been focused on designing SCADA systems with IoT and cloud computing integrated in their architectures.

This is not related with the lack of significance of this kind of integration, but rather with the fact that the problem was mostly treated as a pure engineering project, with a major interest on functionality, security, and connectivity, and with less foundation from a scientific point of view in terms of the optimal design of a SCADA solution that embeds IoT and cloud computing technologies. Table 1 illustrates in a synthetic way the main findings of the literature review. We conclude that the subject of cloud computing and IoT embedded in SCADA systems with application in WWT installations is quite new on the researchers' agenda because only very few contributions have been published until now in journals or conferences with international visibility; thus, we might admit that this technological area still has many hidden, unexplored dimensions.

Table 1. A synthetic view of the related work.

Developments	Limitations	Reference
Platform for real-time water quality monitoring with functionalities for data acquisition, data processing, data communication, data storage, and visualization.	A good candidate for a cloud computing-based SCADA system.	[23]
Experimentation of an IoT based smart water distribution and monitoring system with the integration of fog.	Limited only to functionalities, with no focus from an engineering or scientific point of view on the optimal design of the SCADA system.	[24]

Table 1. *Cont.*

Developments	Limitations	Reference
Integration of IoT technologies in a SCADA system is the central topic of another work.	The focus is more on the description of the technology and introduces a case study that describes the functionalities of a SCADA and IoT system installed for water monitoring purposes.	[25]
Approach for designing a IoT-driven SCADA system in place of remote terminal units (RTUs).	It introduces only initial design steps, as it is explicitly mentioned by its authors.	[26]
Secure fog-based platform for SCADA-based IoT critical infrastructure.	Deploys edge data centers in fog architectures to secure low-latency but with a focus on cybersecurity issues.	[27]

We also conclude that current developments on this topic have been mostly done on system engineering from the angles of functionality, connectivity, interoperability, and in some cases of cybersecurity and modularization, but treated independently, in silos. A step forward is to analyze and tackle the design problem in an integrated manner, and in a way that leads to the capacity of formulating clear arguments for different design decisions. From a performance point of view, it is desirable to balance key target criteria, too. Balancing critical performance characteristics is a proof of a system's maturity from the perspective of the technical systems' evolution [28]. A specificity of our research in relation to this subject is the consideration of a structured methodology for planning and managing the design and optimizing the SCADA architecture with respect to a set of contradicting criteria; thus, positioning our contribution in the field of early design optimization of SCADA systems. We claim that the merit of our work stands in the capacity to argue why a given architecture is proposed and no other alternatives for a given job (e.g., in this case, for the specific situation of WWT installations), why certain technologies are integrated relative to other possible ones, and why certain features are considered to the level of some technologies (both hardware, and software) for building up a disruptive solution.

These conclusions motivated us to investigate new frontiers for designing SCADA systems that embed the latest technological advances for data exchange (IoT, cloud computing). These new frontiers are about capacity: (a) to quantify the impact of each key performance characteristic on the quality of the proposed solution (considering a lifecycle perspective), (b) to solve possible conflicts between some of the performance characteristics or at least to know the limit of compromises (e.g., investment cost versus security, interoperability versus modularity, homogeneity versus adaptability to new technological advances, etc.), and (c) to master complexity during the design process.

4. Methodology

Lifecycle design of a SCADA system falls into the paradigm called the "curse of dimensionality" [29]. We must consider a long list of performance characteristics, that is: secured communication, modularity, interoperability, agility, upgradeability, resilience, scalability, flexibility in operation, efficiency, low cost, homogeneity, integrability, serviceability, easy configuration, redundancy for safety, preventability, architectural reliability, necessary functionality, connectivity, and easy commissioning. Some people might say that several other characteristics cloud be included as well (e.g., replaceability, usability).

This large list of complicated characteristics puts engineers in front of a hard decision. Shall they consider the whole package of design criteria? In the ideal case, this should be the desirable decision, but it comes with a large amount of work for planning and analysis, and with a big number of plausible combinations. This is a big dilemma in engineering in general, not necessarily in this situation.

At this stage of design, optimization cannot use traditional methods such as swarm algorithms, genetic algorithms, or other quantitative optimization tools, because the space of investigation is not a numerical one, either discrete or continuous. As it is shown in the list above, at this stage of design we operate with high level concepts for describing performance. Thus, the key question is: How to reduce the list of characteristics without affecting accuracy of end results? To answer at this question, we use the TRIZ contradiction matrix described in [28]. For simplifying its application, we use a software tool, indigenously developed [30]. Results are illustrated in Figure 2.

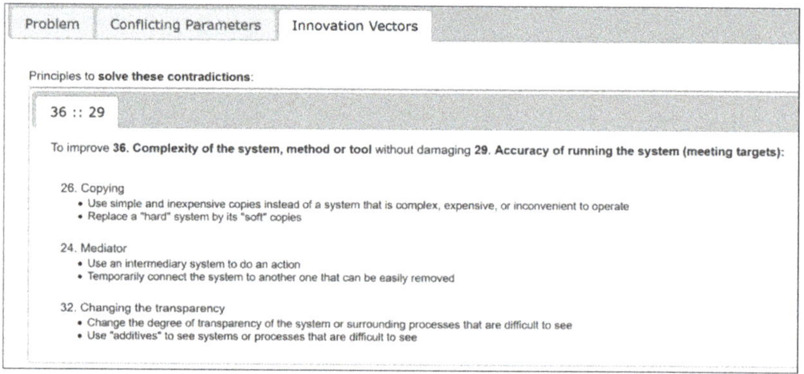

Figure 2. Suggested generic directions of intervention to treat the "curse of dimensionality".

The generic directions of intervention from Figure 2 are further explained. The first direction suggests that the higher the number of software units that will replace hardware units in the architecture of a SCADA system, the better it is for achieving the goal. Additionally, it tells us that we must look for simple and inexpensive solutions to fulfill some functions in the architecture so that replacement and adaptation is not affected as much. The second direction of intervention indicates the presence of an agile "mediator" in the SCADA architecture. Mediation usually happens in conjunction with data exchange between heterogeneous units from the system. For example, how to transfer data from a sensor and an email, or from sensors to an online platform that indicates to city's community the quality of treatment of the waste-water? The last generic direction of intervention highlights the necessity to increase the level of transparency. In other words, it indicates unveiling hidden information about the performance characteristics that ultimately can give us the option to put aside some of these characteristics in the optimization process, and to use them only as checking metrics in conjunction with the proposed solution. If the first two generic indications fall into the category of searching or developing software technologies for solving as many as possible aspects in the SCADA system, the third one encourages us to apply some operations on the list of key characteristics for unveiling aspects that can prioritize them and quantify their impact for giving us the possibility to reduce the list applied for solution optimization. For the first two situations, the problem can be solved with the current technological developments. Thus, at the field level, instead of using simple sensors, we must consider cyber-physical units (CPS), meaning smart sensors, smart actuators, and other smart devices. For example, a smart sensor is a combination of a simple sensor with a microcontroller, which is programable with a high-level programming language (e.g., C) [31]. To ensure flexibility at low costs, a software "broker" that handles heterogeneous devices shall be in place. An investigation reveals that we already have such technologies on the market; they are reliable, and some of them are for free. Such as, for example, the case of node-RED platform from IBM used to wire IoT brokers, hardware devices, and online services [32–34]. Thus, combining the concept of smart units at the field level with a cheap and agile software intermediator (e.g., node-RED), we can quickly and easily design a distributed control SCADA architecture. A distributed control system fits with the goal of designing smart grids of WWT installations. A distributed control system is even better because it facilitates the integration of high cybersecurity paradigms, such as blockchain [35,36] and distributed ledger technologies (DLT) [37] at the edge level (very low level) of the SCADA system [38]. This represents a big step forward for SCADA monitoring and control of WWT units in terms of security.

Just with these simple interventions induced by the first two vectors from Figure 2, we are in the desirable position to eliminate from the design demarche most of the performance characteristics displayed in the initial list; that is, modularity, interoperability, agility, upgradeability, homogeneity, resilience, scalability, integrability, serviceability, easy configuration, efficiency, connectivity, flexibility in operation, and secure communication. They will be used only in the verification stage of the proposed design. The performance characteristic "necessary functionality" is not essentially an element that requires design optimization, meaning it falls into the category of "must requirements", which must be 100% fulfilled. Preventability, one of the performance characteristics from the list, is also not a matter of design optimization. With the capacity to integrate cloud computing in the SCADA system due to smart intermediators such as node-RED, we have the space opened for introducing machine learning algorithms. Due to integrating smart sensors and smart actuators in the SCADA architecture, the collection of relevant data about the operation of system's units is facilitated. Thus, the core interventions in the architecture with the inclusion of CPSs and smart brokers give us the opportunity to perform data analytics for preventive maintenance directly in the cloud. Therefore, we can reduce the list of elements that frames the optimization space at: low cost of the initial investment, redundancy for safety, architectural reliability, and easy commissioning. At this level of complexity, the problem is manageable from the perspective of optimal design. We stress again the aspect that, at this stage of the design, we discuss the qualitative optimization.

The last vector of intervention from Figure 2 suggests the inclusion of "additives" to increase the level of transparency. In this respect, we consider the integration of the AIDA method and the CSDT method in our methodology for planning the design roadmap and visualizing the hidden aspects [39,40].

Figure 3 shows the CSDT planning matrix for the SCADA system. In the planning matrix we consider the following list of generic *inputs* [cloud solution; broker solution; communication solution; local control solution; field solution]. Each *input* has a relative importance rank (*R*), given by engineers to meet the short list of performance characteristics (*outputs*); that is [low cost of the initial investment; redundancy for safety; architectural reliability; easy commissioning]. We gave the ranks for the generic *inputs* with the help of the AHP method [41] by interrogating a focus group of 5 integrators of the SCADA systems. Results are illustrated in Figure 3, column *R*. Each *output* has a relative difficulty rank (*D*) from the perspective of achievement. They have been ranked following the same procedure as in the case of the *R* coefficient. Results are shown in Figure 3, in the raw *D*. To quantify the relationship between *inputs* and *outputs* the following numerical scale is applied: 0 (no relationship); 1 (weak relationship); 3 (medium relationship); 9 (strong relationship); 27 (critical relationship) [42]. Results are shown in Figure 3, in the central matrix of the planning diagram.

The CSDT planning framework also considers several other coefficients. In our analysis the following CSDT coefficients are introduced (see Figure 3): value weight (*W*), technical index of priority (*I*), relative technical effort (*Z*), impact depreciation (*Q*), technical depreciation (*O*), input risk (*J*), difficulty to satisfy inputs (*d*), correlation index of priority (*K*), and input index of priority (*H*). *W* indicates the maximum relative impact that each *output* has within the set of *outputs* to define the optimal design for the SCADA system. *I* coefficient recommends the priority with which each *output* should be tackled. *Z* indicates the relative level of innovation for each *output*. *Q* gives a measure of the negative impact if *outputs* are not well satisfied. *O* is related to the implications on product competitiveness if *inputs* are not well solved. *J* is about the risks associated with poor design of *inputs*.

The coefficient d shows the difficulty to satisfy an *input* by the set of *outputs*. K recommends the priority with which the interdependencies between *outputs* must be analyzed. H recommends the priority with which *inputs* must be satisfied. Figure 3 also puts into evidence the correlations (C) between *outputs*, in the top-right matrix. We consider it superfluous to put all technicalities of the CSDT algorithm here, because they consume a lot of space and do not bring added value to the scope of this research. To find out more details about the calculation of the above-mentioned coefficients, reference [40] should be consulted.

		Easy commissioning	Architectural reliability	Redundancy for safety	Low cost of the initial investment				
PC1	Low cost of the initial investment	−1.62	−1.01	−1.29	<- K				
PC2	Redundancy for safety	−0.6	0.745		−2				
PC3	Architectural reliability	−0.47		2	−1				
PC4	Easy commissioning	C >	−1	−2	−2				
Code	Inputs	$R \backslash D$ 0.31	0.25	0.25	0.19	d	H	J	O
M1	Cloud solution	0.11 9	9	1	27	10.42	94.727	1.1462	5.06
M2	Broker solution	0.25 27	27	9	27	22.5	90	5.625	22.5
M3	Communication solution	0.31 27	27	3	27	21	67.742	6.51	26.04
M4	Local control solution	0.22 9	9	27	27	16.92	76.909	3.7224	15.84
M5	Field solution	0.11 9	9	27	9	13.5	122.73	1.485	5.94
	W	19.08	19.08	12.2	25.02				
	I	61.55	76.32	48.8	131.7				
	Z	5.915	4.77	3.05	4.754				
	Q	25.11	20.25	16.75	22.23				

Figure 3. Planning matrix of the SCADA system.

Information from Figure 3 shows that "Low cost of the initial investment" has the highest impact on the SCADA system ($W = 25.02$), followed by "Easy commissioning" and "Architectural reliability" ($W = 19.08$). At the end of the list is "Redundancy for safety" ($W = 12.2$). This result is logical, even if at first glance it seems counter-intuitive. The capacity to design a reliable SCADA system at low cost places solution providers in the "blue ocean" and disruptive innovation strategy [42], generating a highly competitive advantage.

Figure 3 highlights that the highest priority for best solving the SCADA system is on "Low cost of the initial investment" ($I = 131.7$), followed by "Architectural reliability" ($I = 76.32$), then "Easy commissioning" ($I = 61.55$), and then "Redundancy for safety" ($I = 48.8$). Thus, any effort to identify the cheapest concept for the SCADA system is desirable. Upon that concept, focus must be on defining a reliable architecture in terms of embedded technologies, and then on identifying the business model and operations which lead to an easy commissioning. Once this issue is clarified, the strategy for ensuring redundancy in critical points must be considered.

Results show that the highest level of innovation is required for achieving "Easy commissioning" ($Z = 5.915$), followed by "Architectural reliability" ($Z = 4.77$) and "Low cost of the initial investment" ($Z = 4.75$). According to TRIZ-MC [28], applied with [30], to achieve an easy commissioning without affecting accuracy of the measuring system's performances, we need to adhere to the strategy of making the system sectorial for easy aggregation and disaggregation. To ensure a high reliability without affecting the level of investment, TRIZ-MC [28], applied with [30], leads to the following indication: the need to replace as many as possible hardware units with software units and to increase the local quality (e.g., CPSs).

According to results from Figure 3, the highest competitive problem of a SCADA system (if it is not properly solved) stands on "Easy commissioning" ($Q = 25.11$), closely followed by "Low cost of the initial investment" ($Q = 22.23$). Indeed, the two performance characteristics are strongly connected. The top module of the SCADA system in terms of difficulty to satisfy is the "Broker solution" ($d = 22.5$). With the adoption of latest technology such as node-RED, this challenge can be overpassed.

Priorities to satisfy the generic modules are, according to Figure 3, the following: "Communication solution" ($H = 67.74$) is on the first place, and "Field solution" ($H = 122.73$) is on the last place. There is a logical reason to for this result. Risks about proper operation of the SCADA systems, as well as their competitiveness, mostly count on "Communication solution" ($J = 6.51$; $O = 26.04$) and "Broker solution" ($J = 5.63$; $O = 22.5$). Information provided by the indicators presented in the previous paragraphs are useful to formulate a strategy for designing the SCADA system. We see that the kernel of the design optimization problem is the cluster formed by {"Communication solution", "Broker solution"} against {"Low cost", "Easy commissioning", "Architectural reliability"}.

CSDT introduces a series of vectors of innovation for tackling the conflicts between various performance characteristics. They are generated by TRIZ-MC [28], applied for solving contradictions between various performance characteristics. According to data in Figure 3, the order of tackling contradictions, given by the indicator K, is: (1) PC1 against PC4 ($K = -1.62$); (2) PC1 against PC2 ($K = -1.29$); (3) PC1 against PC3 ($K = -1.01$); (4) PC2 against PC4 ($K = -0.6$); (5) PC3 against PC4 ($K = -0.47$) (note: the sign "$-$" indicates the negative correlation, or contradiction). We use the software tool available at [30] to reveal the vectors of innovation. They are presented in Table 2.

Table 2. Generic vectors of innovation to tackle contradictions.

Conflict	Vector of Innovation	Code
PC1 vs. PC4	Replace hard parts of the system with reconfigurable modules	V1
	Make the system sectional (for easy aggregation or/and disaggregation)	V2
	Change the degree of flexibility	V3
	Replace the system with more inexpensive units, comprising properties	V4
PC1 vs. PC2	Each part to be placed under the most favorable conditions for operation	V5
	Change the degree of flexibility	V6
	Replace a homogeneous system with a composite one	V7
	Introduce a neutral element	V8
PC1 vs. PC3	Use actions that make the system resonate	V9
	Make such different parts of the system to carry out different functions	V10
	Replace hardware with software	V11
	Replace a homogeneous system with a composite one	V12
PC2 vs. PC4	Replace the system with more inexpensive units, comprising properties	V13
	Temporarily use an intermediary system to do an action	V14
	Replace a homogeneous system with a composite one	V15
	Extract some functions or units and consider a way around	V16
PC3 vs. PC4	Compensate for low reliability with countermeasures in advance	V17
	Increase the degree of system's segmentation	V18

The large number of combinations that can be generated with the vectors from Table 2 is not desirable from a practical point of view. Therefore, we consider the AIDA method [39] to select the best combinations; that is, the best instance for each of the five clusters from Table 2. To select the best vector for each decision area (column "Conflict" in Table 2), the AIDA method operates with a list of constrains, compared to which every vector is analyzed. To relate with the SCADA system design, we have selected the following list of constrains [(c1) the possibility to make the whole commissioning by a single integrator; (c2) less or no sensitivity to various fields of technologies; (c3) easy to create connectivity between objects; (c4) easy to program].

The application of the AIDA method for the list of vectors from Table 2 is shown in Figure 4. For each decision area, the AIDA method establishes the level of influence of each element from the row on each element from the column (1, 2, or 3). Then, the value from every box from matrix one is multiplied with the value from the corresponding box from matrix two and introduced in matrix three. The average of values on columns and rows is also introduced in matrix three. For example, in the case of vector V1 (see Figure 4, top left corner), values in the third matrix along the V1 column are: $9 = 3 \times 3$; $6 = 3 \times 2$; $6 = 3 \times 2$; $3 = 1 \times 3$, and $6 = $ sum $(9, 6, 6, 3)/4$. For each decision area, the vector with the highest average value is selected from the list of candidate vectors from Table 2 (see Figure 4). Table 3 shows the selected vectors of innovation for each decision area after the application of the AIDA method.

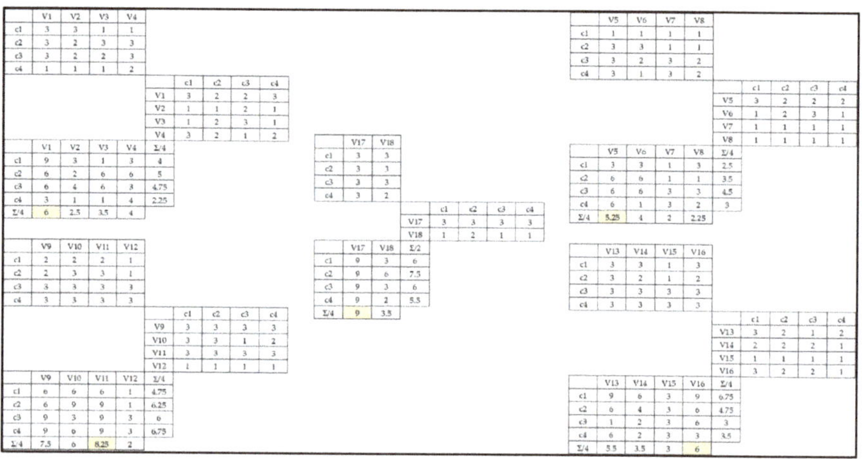

Figure 4. Results of the AIDA method application.

Table 3. Selected vectors of innovation.

Conflict	Vector of Innovation	Code
PC1 vs. PC4	Replace hard parts of the system with reconfigurable modules	V1
PC1 vs. PC2	Each part to be placed under the most favorable conditions for operation	V5
PC1 vs. PC3	Replace hardware with software	V11
PC2 vs. PC4	Extract some functions or units and consider a way around	V16
PC3 vs. PC4	Compensate for low reliability with countermeasures in advance	V17

To manage the design process of the SCADA system, the CSDT method provides a management flow that takes information from the planning matrix (Figure 3). The algorithm for defining the design flow is a bit laborious and can be consulted in [40]. However, it is less important for the scope of this research. What matters here is the result; that is, the design flow. In principle, the design problem is divided into very elementary steps, following a certain rule to focus on different aspects of the design problem. In this way, a complex case (with many possible patterns of evolution) is brought to a level of analysis that is manageable by engineers.

To visualize the design flow, some conventions are used. They are further introduced: (1) symbol "<>" indicates a link between two subsequent steps from the flow; (2) the symbol "&" describes the request to analyze the correlation between two outputs; (3) the symbol "|" asks to apply a given vector of innovation (see Table 3) to solve a negative correlation between two outputs; (4) the symbol "—" represents the process of conceptualizing, finding a partial or complete solution for a given input with respect to a given output or a pair of outputs. With these clarifications, in the next paragraph we introduce the evolutionary design flow for the SCADA system:

- Flow 1: M3—(PC1&PC4) | V1 <> M4—(PC1&PC4) | V1 <> M2—(PC1&PC4) | V1 <> M1—(PC1&PC4) | V1 <> M5—(PC1&PC4) | V1
- Flow 2: M3—(PC2&PC1) | V5 <> M4—(PC2&PC1) | V5 <> M2—(PC2&PC1) | V5 <> M1—(PC2&PC1) | V5 <> M5—(PC2&PC1) | V5
- Flow 3: M3—(PC3&PC1) | V11 <> M4—(PC3&PC1) | V11<> M2—(PC3&PC1) | V11<> M1—(PC3&PC1) | V11<> M5—(PC3&PC1) | V11
- Flow 4: M3—(PC4&PC2) | V16 <> M4—(PC4&PC2) | V16 <> M2—(PC4&PC2) | V16 <> M1—(PC4&PC2) | V16 <> M5—(PC4&PC2) | V16
- Flow 5: M3—(PC4&PC3) | V17 <> M4—(PC4&PC3) | V17 <> M2—(PC4&PC3) | V17 <> M1—(PC4&PC3) | V17 <> M5—(PC4&PC3) | V17
- Flow 6: M3—PC3 <> M3—PC2 <> M4—PC2 <> M4—PC3 <> M2—PC3 <> M2—PC2 <> M1—PC3 <> M1—PC2 <> M5—PC2 <> M5—PC3

The design flow comprises six sub-flows. The last sub-flow looks a bit different than the others because it treats the design with respect to two performance characteristics that are positive correlated; thus, there is no need to provoke a resolution of conflicts. The six sub-flows above presented can be tackled in more cycles (e.g., 2 or 3 cycles) if the results are not mature after the first cycle. As the flows show, the conceptualization (design) process of the SCADA system is divided into 35 elementary steps. At each increment, an elementary problem is analyzed. Thus, complexity is better administrated. As can be seen, the design process is an evolutionary one and provides a structured space for search and ideation. With this algorithm, the chances of omitting important aspects of design are significantly lowered. The whole methodological effort finally leads to the evolutionary design flow, which is effectively applied in practice to formulate a concrete solution for the SCADA system. The algorithm does not restrict the ideation space, but rather focuses ideation to the right directions. This issue is essential for avoiding trial-and-errors models, which are time consuming and usually lead to solutions that embed many compromises. Following the flow of the algorithm, some steps can be simply solved by searching onto the market for existing technologies that satisfy the request; but other steps indicate the need for new developments. This makes a leap forward to current practices of integrators that mostly count on integration of current technologies, with no focus on research and development that, in many cases, leads to suboptimal solutions with a lot of drawbacks, including the setup costs and lifecycle costs.

5. Case Study and Results

Our ambition through this case study is to propose a solution that overcomes some of the identified drawbacks of actual implementations of SCADA systems (see Section "Background"). Moreover, the ambition is also to define the SCADA system with an eye on the set of performance characteristics that describes the lifecycle perspective of the system. These performance characteristics are nominated in the first paragraph of the Section "Methodology". They will be used as checklist for the proposed solution, knowing that by applying the lines of evolution indicated in Figure 2, we propose a methodology for qualitative optimal design that escapes from the trap of the "curse of dimensionality". The design flow introduced in the end part of the Section "Methodology" is further applied to design a disruptive, lifecycle-oriented solution for the SCADA system. As it will be seen, the solution proposed at the end of this process embeds cloud computing and data analytics, with significant advantages in terms of applying machine learning algorithms for preventive maintenance and for providing very valuable information about lifecycle behavior of the SCADA system that enables the improvement of future designs.

It is superfluous to present in the paper each of the 35 steps of the design flow, because this is not the focal point of the research. It is more important to indicate the generic results that came out of traversing the design flow, and rather as exemplification to illustrate some steps.

5.1. Main Outcomes from the Application of the Design Flow

The first step of the evolutionary algorithm from the methodology is: "M3—(PC1&PC4) | V1", which is translated as: define "communication solution" to minimize "initial investment costs" and to maximize "easy commissioning", considering in the resolution of this job the guiding vector "replace hard parts of the system with reconfigurable modules". A reconfigurable module is a module that embeds the following properties: scalability, convertibility, modularity, flexibility, and integrability. The current solutions of communication include many intermediary hardware modules at the level of communication network (e.g., radio, GPRS, satellite, VPN, Internet). A solution to reduce complications with these technologies is to replace them with IoT and cloud computing. IoT is modular, flexible, integrable, convertible, and scalable; thus reconfigurable. By adopting this type of communication, costs with integrators are drastically reduced. Commissioning is sharply simplified, because most of the work with setting up and programming the control logic can be done without locating specialized personnel in the field.

The second step of the evolutionary algorithm is: "M4—(PC1&PC4) | V1", which means: define "local control solution" to minimize "initial investment costs" and to maximize "easy commissioning", considering in the resolution of this job the guiding vector "replace hard parts of the system with reconfigurable modules". Searching on the internet with the keywords that describe reconfigurability (see the above paragraph) on control units, we have identified several researches and technologies in the paradigm "master-slave", with reconfigurable master unit, and with CPSs (e.g., smart sensors, smart actuators, etc.). A special attention was assigned to industrial technologies that respect these properties, for the main reason of high reliability in intensive industrial tasks. Among these technologies we have identified the presence of modular PLCs units that can be easily configured for distributed local networks and can be linked with no effort to IoT-based communication solutions.

We followed the next steps of the evolutionary algorithm in the same manner as the first two presented above. It does not make much sense to put all these steps here, because this process does not reflect the focus of the paper. However, four more steps from the algorithm have been selected for presentation to highlight the way we addressed the other four key vectors of innovation. In this respect, for exemplification, we introduce here the steps: "M2—(PC2&PC1) | V5", "M1—(PC3&PC1) | V11", "M5—(PC4&PC2) | V16", "M2—(PC4&PC3) | V17".

"M2—(PC2&PC1) | V5" is formulated as: define "broker solution" to increase "redundancy for safety" without affecting "costs of initial investment", considering in the resolution of this issue the principle "each part to be placed under the most favorable conditions for operation". In practice, redundancy is usually about doubling some elements in the system such that, in the case that one of them is down, the other one enters into action to compensate the temporary unavailability of the first one. Redundancy for safety on the broker's side is defined by the architecture of the MQTT broker. Indication "each part to be placed under the most favorable conditions for operation" leads us to the idea of setting up a cluster environment for high availability of MQTT. Searching for existent technologies onto the market, we have identified that HiveMQ provides such facility without involving additional costs, just a special configuration of a cluster of brokers. In the cluster, a message only gets forwarded to other cluster nodes if a cluster node is interested in it. This permits to build a big cluster with a lot of nodes because it reduces the network traffic, and this prevents nodes from forwarding unnecessary messages. In addition, cluster subscriptions work dynamically. As soon as a client on a node subscribes to a topic it becomes known within the cluster. If one of the clients from the cluster publishes to this topic, the message will be delivered to its subscriber no matter to which cluster node it is connected. HiveMQ uses a message serialization mechanism to share publishes between cluster nodes, which significantly reduces the network traffic between nodes. When a cluster node where a client subscribes goes down, we need to have an advanced logic such as the client to connect to another node. To avoid the problem of reconnecting on the client side, the traditional solution is to have an advanced reconnect logic where the client connects to another cluster node, but the fallback IP could be hard-coded, and this is not desirable. Therefore, the innovation vector "each part to be placed under the most favorable conditions for operation" suggests the use of a "load balancer", which permits the construction of a high availability environment without the need of implementing any advanced reconnecting logic on the client side. This approach can be easily used with devices that are already deployed in the field, with no need to change their already configured connection information. This solution avoids additional costs to configure and commission the SCADA system, as well as requiring less skills and dramatically reduces the implementation time.

The design step "M1—(PC3&PC1) | V11" looks like this: define the "cloud solution" such that to improve "architectural reliability" without increasing the "investment costs", considering the guiding vector "replace hardware with software". This indicates to operate "soft" interventions. To increase architectural reliability from the cloud perspective, we must avoid various potential failures in cloud computing systems. To reduce initial investment costs, the solution is to use the cloud services of a third party. Thus, reliability issues must be treated with the cloud provider. They are related with availability and resilience. Cloud platforms are intrinsically built to tolerate failures and provide features to help build reliable and highly available systems. High availability is usually accomplished with redundant deployments of the system, meaning that two or more identical application instances run simultaneously (or only one instance is active or primary at a time, and the second one on standby). To make the solution resilient, it is necessary to design the application that runs on the cloud as highly decoupled microservices (i.e., small services focused on performing a specific business function). This design enables a resilient behavior, meaning that when one service fails, the others can continue working. It also allows continuous delivery, with each service updated and deployed independently without interruptions.

In the case of the design step "M5—(PC4&PC2) | V16", the challenge is to define "field solution" so that to have an "easy commissioning" without affecting "redundancy for safety". To tackle the challenge, the core vector of innovation is "extract some functions or units and consider a way around". Field solution is referring to the field devices, augmented with intelligence and communication capabilities. Redundancy for safety is not referring to doubling devices, which can be done for very critical aspects, as it is happening in aviation. In this case, redundancy for safety is looking for avoiding the breakdown of communication between field devices and the local control unit. To be aligned with "extract some functions or units and consider a way around", the solution is to analyze both the communication protocol and connections, including contacts, etc. Operating with CPSs might generate a problem from this point of view, therefore we have investigated what possibilities exist to the level of PLCs for ensuring local fieldbus communication in a safe way. We have identified solutions to easily create redundancy in the local communication network, but this requires at least two PLCs as slaves and one as master. Usually, in a WWT installation or plant there are several PLCs installed for the local control; thus, this strategy is plausible. In terms of fieldbus communication, PROFIBUS supports the before mentioned strategy.

For the last design step exemplified in the paper, "M2—(PC4&PC3) | V17", the goal is to define a "broker solution" such that to have a good "architectural reliability" without complicating "commissioning". To tackle this job, the vector of innovation is "compensate possible low reliability with countermeasures in advance". In a paragraph before we have already mentioned about the solution proposed for "broker". It has a high reliability; however, a supra-control is very useful if it can be done without affecting the effort for commissioning. The solution aligned with the vector of innovation is the inclusion of a flow-based tool that connects hardware, services, etc. There are such tools available, easily programmable with visual objects, reliable, and some of them are for free (e.g., node-RED).

The final selected solution for the new SCADA architecture is illustrated in Figure 5. It also visualizes the selected technologies we have decided to integrate within the validation workbench. Details on these aspects is provided in the following sections of the paper.

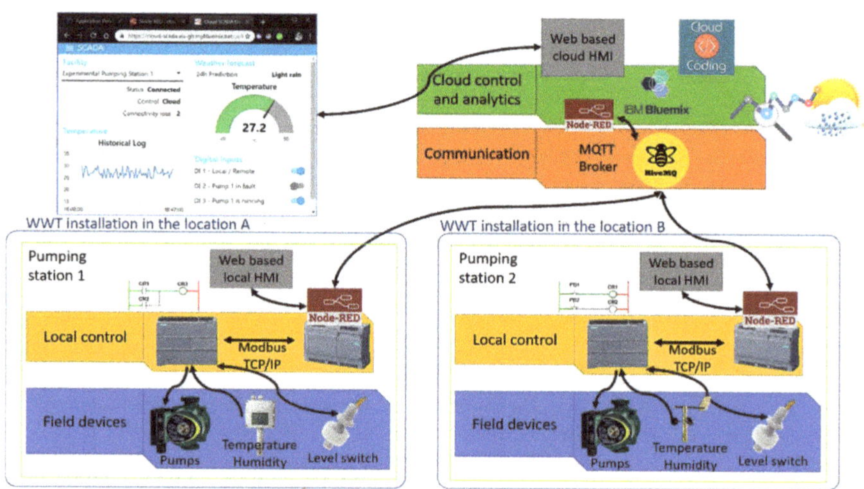

Figure 5. Exemplification of the distributed architecture for the new SCADA system concept.

However, several other technologies for the IoT broker, API platform, local control units, and cloud computing exist on the market. This means that the proposed architecture can also be built with alternative technologies. For example, Allen Bradley can be a reliable alternative to Siemens PLCs and IoT gateway. Azure or AWS can anytime be an alternative to IBM cloud, but also any private cloud can replace the proposed one. For some reasons (e.g., latency), local private clouds are preferable. However, this comes with a high investment, therefore the wise approach is to share a private cloud resource with some other local enterprises, or to rent local cloud resources from providers located in the geographical proximity. The HiveMQ broker can be replaced with Waterstream or Kafka without affecting system's performances. Flow, n8n.io, ioBroker, ThingsBoard.io, iot-dashboard, AWS IoT, and other platforms are a good alternative to node-RED for interfacing heterogeneous objects and services.

5.2. Validation Workbench

As a validator, the case study takes the generic outcomes from Section 5.1 and applies them to an experimental workbench of a technological unit from WWT industry, namely, a pumping station. The pumping station has two pumps, and it is controlled by a programmable logic controller (PLC) [43]. Two level switches are used to detect the water level in the tank. The PLC is included as part of the CPS concept, but with a clear advantage in terms of resilience and safety, meaning that, in the case of some accidents (outside the control capacity of the operator) the internet connectivity is down, the logic embedded in the PLC can run the station safely, until the connection is re-established. For communication, an industrial certified IoT gateway is used to exchange data between the pumping station control unit and the cloud service via a Message Queuing Telemetry Transport (MQTT) protocol broker [44]. The architecture of the proposed experimental workbench for a WWT installation is presented in Figure 6.

The complete pumping station control logic and data analytics runs in the cloud platform and the required actions are transferred to the pumping station at specific time intervals. In the case of connectivity-loss between the PLC and the cloud service, a local control logic that is implemented in the PLC takes over the control, as it was already mentioned in a previous paragraph. The local control logic is designed to provide a set of minimum requirements considering technological functionalities of the pumping station. When the connectivity with the cloud platform is re-established, control of the pumping station is turned back to the control logic that runs on the cloud platform.

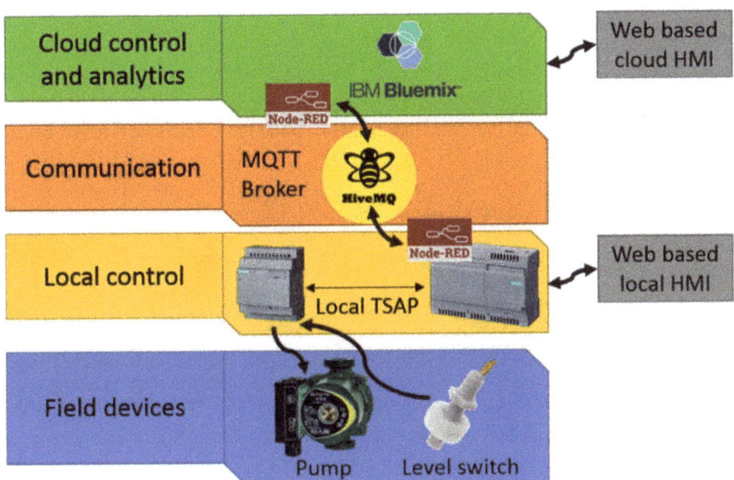

Figure 6. Architecture of the SCADA system for the validation workbench.

A local Transport Services Access Point (TSAP) is established between the local PLC and the IoT gateway. In the experimental workbench we have installed Siemens's technology, which provides reliable and safe local communication and compatible interface units between PLC and web. The use of node-RED programming tool creates the possibility to easily design (i.e., using visual objects) both the environment for wiring together hardware devices, APIs, and online services, as well as to design with visual objects the local Human-Machine Interface (HMI) that is accessible from browser. This offers a huge advantage in terms of accessibility and flexibility for monitoring the WWT process. As an IoT broker, we have selected one of the most reliable solutions from the market, specifically the HiveMQ [45]. A comparison of various MQTT technologies can be found at [46]. Thus, alternatively to our choice, some other valuable technologies are already available. Our decision to select HiveMQ stands in the fact that this IoT broker is designed for cloud native deployments to make optimal use of cloud resources. MQTT for HiveMQ reduces network bandwidth required for moving data; thus, it leads to efficient IoT solutions that lower the total costs of operation over the lifecycle. Additionally, HiveMQ connects any device and backend system in a reliable and secure manner via the IoT standard protocol MQTT and ensures a quick send and receive of data from connected devices (a key safety aspect for WWT installations). HiveMQ is also architected for scalability and reliability and uses industrial standards to reduce the risk of losing data. The open API of this broker allows flexible integration of IoT data into enterprise systems and pre-built extensions for quick integration to other enterprise systems such as Kafka, SQL, and NoSQL databases. Finally, it has a multi-cloud strategy that allows the MQTT broker to be deployed on private, hybrid, and public clouds. This represents a big advantage in terms of agility, redundancy, and flexibility over lifecycle. For cloud computing, in this experimental workbench we opted for IBM Bluemix (now branded IBM Cloud) [34,47]. It offers a web-based cloud HMI; and with node-RED it can be easily connected to the IoT broker. Our option for this cloud technology was influenced by a series of advantages, such as the possibility to merge public, private, hybrid and multi-cloud, and inclusion of AI algorithms for fast and easy data analytics, without the need of advanced programming skills, but also because it is easy to use and is unified across multiple deployments and supports mission critical workloads.

There are several other advantages of the proposed SCADA control architecture compared to the traditional way of controlling technological functionalities related to the WWT industry. Some of the most important ones are further highlighted:

- We can reduce the costs related to local control unit (e.g., work memory) since the control logic runs in cloud and only reduced functionalities are implemented in the PLC. Therefore, we do not need to consider high-performance control units for local control; simpler and cheaper models of PLCs are sufficient to fulfill the job.
- Adding or removing functionalities in the cloud control logic is much easier and it can be done without being physically present in the field; thus, lowering the operational costs over lifecycle and initial investment costs. This ends with reduced commissioning and adjustments costs with solution integrators and control units.
- Commissioning can be done remotely, without involvement of experienced personnel in the field. This is done by analyzing data received on the cloud platform and adjusting the control logic accordingly.
- The architecture is highly reliable and scalable. Similar control functions and data processing algorithms can be used for geographically distributed technological units, providing repeatable control logic results. If adjustments are required after deployment, they can be easily done in the control logic developed on the cloud to redefine all technological functions of WWT installations without additional field actions.
- There is no need to pay static IPs to communication technology providers in order to access the intelligent gateway; thus, reducing the lifecycle costs of the system.
- The control logic that runs on the cloud computing engine can be correlated with other services. For example, meteorological information and alerts can trigger the process of emptying a pumping station to preserve bacteria concertation for technological process reactions within specific limits or to provide the required inlet capacity of the sewage system.
- The know-how for controlling the technological process stays in the property of the utility company; thus, avoiding captivity to some integrators.

The subsequent sections provide a more detailed description of how the most important components of the local control units were configured and how they work.

5.2.1. Local Control Unit

The local control unit has a Logo8! PLC manufactured by Siemens (Munich, Germany), which controls the two pumps and reads the status at the level of switches. An IOT2040 IoT gateway manufactured by Siemens is used to read or write data to the PLC by means of a communication protocol and exchange data with the cloud control platform [44].

When local control is selected or when the connectivity with the cloud computing platform is not available because the internet connection is down, the local control unit uses the local control logic that is designed so to provide the minimum technological requirements with respect to the operating and technological modes of the station. Three operating modes and two technological modes of the pumping station were implemented by software means and loaded into the local control unit.

The first operating mode uses one pump while the second pump is used just in case the first pump requires maintenance or is tripped due to overload or overcurrent. The second operating mode uses both pumps at the same time. The third operating mode periodically alternates the two pumps to balance the workload of the pumps. In the empty technological mode, the pumping station starts if the water in the tank reaches the second level switch and stops if the water reaches the first level switch. In the fill mode, the pumping station starts if the first level switch is reached and stops if the water reaches the second level switch.

These operating and technological modes were considered due to the multitude of uncertainties experienced by the authors in the field of WWT industry. Mainly, uncertainties are generated by technological design mistakes or major changes in water characteristics (e.g., water flow, water requirement).

Siemens IOT2040 is enhanced with a Yocto Linux distribution as the operating system [44]. On top of the operating system the Node-RED programming and runtime environment, with specific functions or nodes, is installed [32].

5.2.2. TSAP Connectivity

The IoT gateway uses a special developed library named *node-red-contrib-s7* to communicate with Siemens PLCs [32]. A software method to connect to Logo8! PLCs considers the end-point communication protocol named Transport Service Access Point (TSAP) which defines the channel used for the communication and the number of the communication modules between the server and the client on top of local network (TCP/IP).

In this experimental workbench, the PLC is configured as a Server with its own IP address, and it can be accessed by a Client with the configured IP address using the channel with the number 20. Since the PLC comes with internal communication functionalities, the number of the communication module is 00. Therefore, the PLC TSAP is 20.00. Depending on the equipment that accesses the server, TSAP configuration must be changed accordingly. The TSAP for the client is configured to 03.80, as seen in Figure 7.

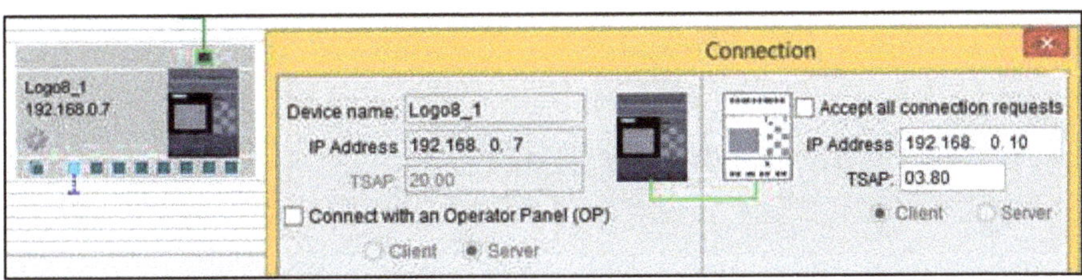

Figure 7. TSAP configuration on the PLC side.

On the node-RED application side, which runs on the IOT2040 IoT gateway, the S7 in operation was configured as the TSAP configuration for PLC. Additionally, the IP of the PLC, the cycle time for reading data and connection timeout were configured, as it is seen in Figure 8.

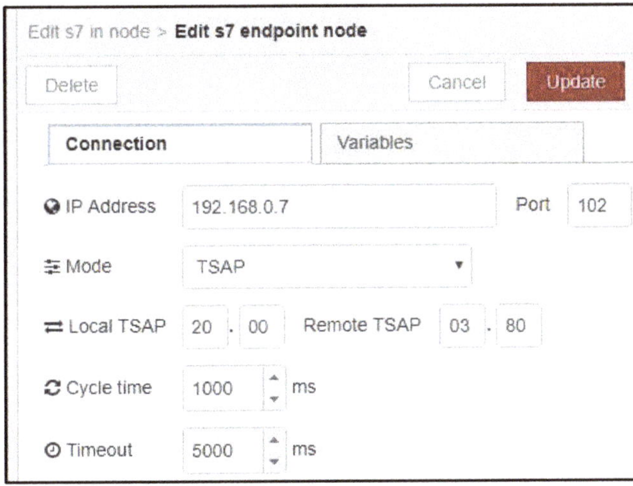

Figure 8. TSAP configuration on the node-RED side.

Writing or reading data, inputs, and outputs in Logo8! PLC is achieved by means of *s7 in* and *s7 out* node-RED functions [32] with respect to mapping resources in the memory of the PLC and the variable memory table (see Figures 9 and 10).

Device Type	VM (From)	VM (To)	Range
I	1024	1031	8 Bytes
AI	1032	1063	32 Bytes
Q	1064	1071	8 Bytes
AQ	1072	1103	32 Bytes
M	1104	1117	14 Bytes
AM	1118	1245	128 Bytes
NI	1246	1261	16 Bytes
NAI	1262	1389	128 Bytes
NQ	1390	1405	16 Bytes
NAQ	1406	1469	64 Bytes

Figure 9. PLC mapping resources of Logo8! PLCs.

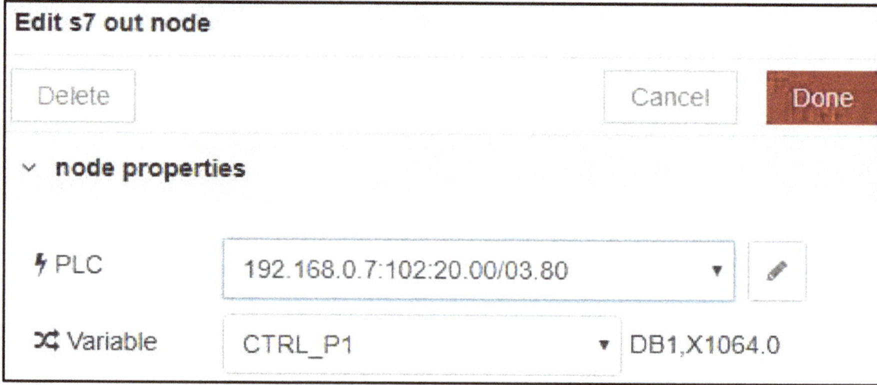

Figure 10. Writing the first digital output of Logo8! PLC.

Therefore, the node-RED application that runs on the IOT2040 to read the first digital input of the PLC needs to access the memory zone *DB1,X1024.0*. While writing the first digital output of the PLC, the memory zone that must be accessed is *DB1,X1064.0*, as it is seen in Figure 10.

5.2.3. User Interface

Figure 11 presents one example of the local user interface developed to monitor and control the functionality of the pumping station. Development of such a user interface requires specific dashboard functionalities to be installed in the node-RED environment. The *node-red-dashboard* library is used to develop the user interface, presented in Figure 11. The user interface is divided into several sections: pumping system status, pumping system settings, digital inputs, digital outputs, parameters of pump 1 and parameters of pump 2.

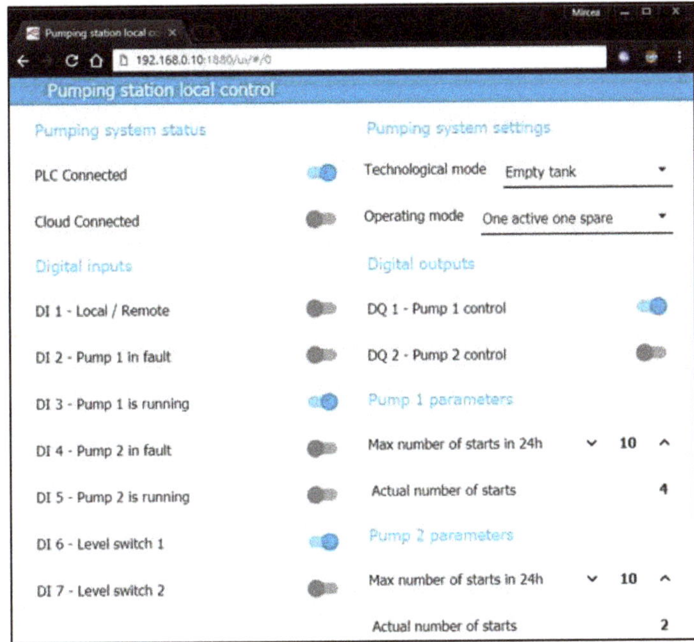

Figure 11. Local user interface.

The pumping system's status section provides information whether the IOT2040 was able to connect to the local PLC control unit and to the cloud platform. To provide this functionality, the IOT2040 gateway uses watchdog variables. The watchdog variables values are incremented every 5 s. If the sent value is not received back by the IOT2040 gateway within 5 s, the connectivity with the PLC or cloud platform is considered lost.

If the "local control mode" is selected or if there is no connection with the cloud platform, pumping system settings allow the local control user to change the technological mode and operating mode of the pumping station.

If the cloud platform is connected to the local user interface, it can monitor the settings selected by the application running in the cloud.

Digital inputs section provides information related to the status of digital inputs. In Figure 11, it can be observed that the local switch selector is switched "on" for local control; thus, pump 1 runs (i.e., feedback is received from auxiliary contacts of the relay that drives the pump) and the water in the tank is above the first level switch.

The digital outputs section provides information related to the status of the digital outputs. As in Figure 11, it can be observed that the PLC has activated digital output 1, which is connected to the relay coil that supplies with energy the first pump.

The parameter sections of pumps 1 and 2 provide information related to the maximum number of starts in 24 h, configured for each pump, and the actual number of starts. If the maximum number of starts of a pump is reached, the pump's control is deactivated to protect that pump from too many starts. Usually, the maximum number of starts is mentioned by the manufacturer (i.e., in the pump datasheet).

User interface can be locally accessed by means of a PC or a mobile phone via a network cable or a Wi-Fi access point without the need of other proprietary software, just by introducing the IP address of the IOT2040, followed by the port and the path to the user interface page: *192.168.0.10:1880/ui/#/0*. This is an important feature for flexibility, at no costs. All modern browsers must provide the required framework to access the web page that was created in the IOT2040 gateway. A user ID and password can be configured to provide security and allow the authorized staff to configure, monitor, and control the pumping station.

5.2.4. Data Exchange over MQTT

Using the TSAP communication protocol and the *s7 in* and *s7 out* node-RED functions, data are read or written from and to PLC by the IOT2040 gateway. To exchange data between the IOT2040 gateway and the cloud platform, the MQTT communication protocol is included in the SCADA architecture. On top of being an asynchronous communication protocol, MQTT has the advantage of having a lightweight data framework and does not require high computational resources [35].

The MQTT uses a publish and subscribe mechanism with several qualities of data transfer checks. The node-RED programming environment comes with *MQTT in* and *MQTT out* functions already installed [33]. These functions need to be configured to work with an MQTT broker. The HiveMQ MQTT broker is used by the IOT2040 gateway and cloud platform to publish or subscribe to specific topics.

Figure 12 presents the configuration of *MQTT out* function in the node-RED programming environment that runs on the IOT2040 gateway. The node-RED application that runs on the IOT2040 gateway publishes data under the topic *PumpingStation1/LocalToCloud* with a JSON structure, where data representation is described (see Figure 13).

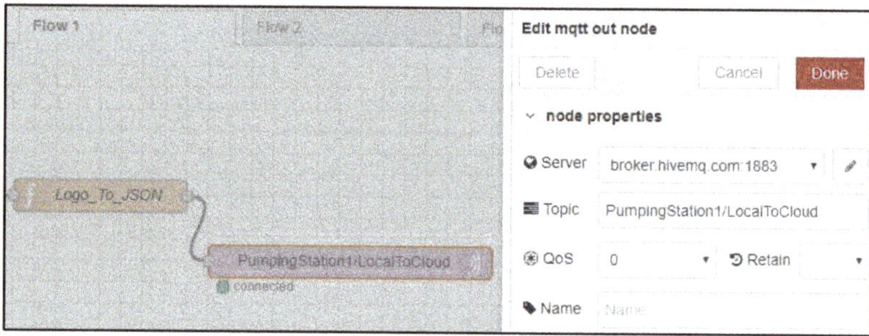

Figure 12. node-RED *MQTT out* function configuration.

```
 1   {
 2       "PumpingStation1": {
 3         "PLC": [
 4           {
 5               "Tag": "Digital Input 1",
 6               "value": True or False,
 7           },
 8           . . .
 9           {
10               "Tag": "Digital Input 7",
11               "value": True or False,
12           }
13         ]
14       }
15   }
```

Figure 13. JSON format of digital inputs representation.

Cloud platform application control logic has an MQTT node that is subscribed for the above-mentioned topic and receives from the MQTT broker the messages under subscribed topics. In a similar way, the cloud platform publishes data to the topic *PumpingStation1/CloudToLocal* and the IOT2040 modules read the data by subscribing to the topic and use the information accordingly.

5.2.5. Cloud Computing Platform

The IBM Bluemix (IBM Cloud) cloud computing platform was considered as the development and running environment of the remote monitoring and control application for the testbench. Registration for a lite account provides access to a specific runtime and storage memory in the cloud and free data exchange with real-world devices up to a bandwidth limit is ensured [34]. For a real WWT plant, a professional account must be registered for an affordable annual fee; thus, being offered adequate resources in the cloud. The result of the configuration process is seen in Figure 14.

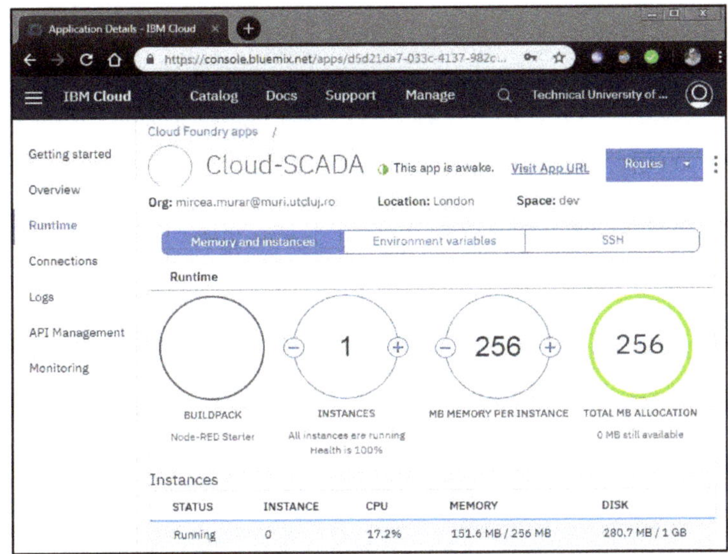

Figure 14. Result of the configuration process of IBM Bluemix environment.

Another instance of node-RED programming and visualization environment was installed and prepared for operation on top of the IBM Bluemix infrastructure along with specific functionalities required to exchange data with the node-RED application, which runs in the IOT2040 gateway of the testbench. The applications developed in the two instances of node-RED transfer data from PLC to the cloud platform and vice-versa by subscribing and publishing in specific topics of an MQTT broker.

Cloud control can be enabled or disabled from the local HMI. If the connectivity between the pumping station and the application that runs in the cloud is established, pumping station's control is handled to the application logic which runs in the cloud. The node-RED application, which runs on the cloud platform, receives information from its local control counterpart application, which is further analyzed, and results are sent back to the control unit.

In a more advanced application, the HMI that runs in the cloud must have several sections, one for every remote-controlled automation system. Since the testbench has only one remote automation unit, the HMI that runs in the cloud is able to provide monitoring and controlling functions related to the pumping station. Therefore, the interface is similar to the one that runs locally on the IOT2040 gateway (see Figures 5 and 11).

6. Discussion

Discussions from this section are grouped into two parts. The first part provides information upon the performed tests on the experimental workbench. The second part relates the results with the performance characteristics highlighted in the methodology.

Tests we have run on the proposed cloud-based SCADA system indicated a proper behavior. Non-critical time lags were observed after the HMI applications were put into operation on both local and cloud HMI. We assume that fine tuning and making other adjustments on the operating system settings, on IOT2040 and node-RED application, might fix this drawback. Data on the calibration process are not provided in this paper. In addition, we conclude that ensuring a high-speed internet connection and running the cloud application on a local cloud or on the company cloud computing platform would deliver better time response than the one observed on IBM Bluemix. This option depends on several variables, such as the existence of the local cloud, as well as the criticality of time lags from the perspective of the WWT installation's operator.

We investigated here disruptive technologies, which are at their primary stages of evolution on the S-curve. Thus, we recommend this SCADA architecture for non-critical WWT installations, not for the premium ones. The single reason is the commonsense prudence until the long-term reliability of these disruptive technologies is not fully proven. Of course, this takes time, but we can conclude that cloud computing and IoT driven SCADA systems represent the future. So, we recommend not using yet this type of technologies on critical WWT processes (e.g., chemicals dosing [48]) before testing them into non-critical areas, including other industrial processes than WWT, because it can provide a more efficient and cost-effective way of controlling and monitoring specific processes in many industries and provide novel and effective approaches to solve problems and adjust control logic [49].

Another aspect to comment is the friendliness of the user interface. Operators of traditional SCADA systems are used with a specific graphics. On the web, these graphics are not as pleasant as on a desktop. However, with an additional effort these graphics can be achieved on web, too. However, this increases the initial cost, because it requires skilled people to program in JavaScript and HTML for releasing a professional HMI in both local and cloud applications, similar with the ones that ordinary operators are familiar with in the actual SCADA systems.

In addition, we conclude that cloud computing and IoT can provide lots of advantages since the applications that run on cloud computing platforms can be connected to other services in distributed networks over the Internet; thus, improving efficiency, flexibility, agility, and rapidity in interventions of the controlled process, while reducing the lifecycle costs. They also have good security features. Nevertheless, cybersecurity of IoT devices is at an early stage of development, therefore the use of these disruptive technologies for critical installations should consider this delicate aspect. More and more researches in this field are already reported, and new start-ups are working on it; thus, we expect that security performances will soon reach an acceptable level also on this type of data exchange technologies [27]. One might accept that cybersecurity is a never-ending issue, hence this could be a direction for future researches related to WWT installations. However, the use of cloud computing facilitates the integration of machine learning and deep learning for AI-driven cybersecurity, which is a step forward in this technological field. Application of AI models for data analytics is another future area of investigation, especially from the perspective of producers of the constitutive technologies of a SCADA system to support their demarch on product-service innovation [49]. Reliability, in terms of not losing data during transfer over long distances via Internet, is another area that necessitates future researches, because measurements we have performed at the level of currents in the network indicate that enforcement, buffering and other types of redundant or backup architectures might be useful in this respect.

A systematic analysis of the performances achieved by the proposed SCADA architecture with respect to the pool of objective functions related to a lifecycle orientation (introduced at the beginning in the section "Methodology") is done in Table 4.

Table 4. Analysis of the proposed solution against various key performance indicators.

Performance	Explanations
Modularity	Solution embeds capacities to be developed in distributed architectures using modular units
Interoperability	System is capable of connecting heterogeneous devices and services and to ensure communication between them
Agility	As designed, the system can be scaled-up or down locally and as a whole, according to the specific needs in any moment over the lifecycle of the system
Upgradeability	Any unit of the system can be replaced without complications and there is no major barrier to update the system with the latest technologies at every layer of the architecture, from the field to the cloud level
Resilience	Because of the selected technologies, the system encompasses capacity to be rapidly put into operation after any lockdown, and the proposed architecture permits local operation even if the connection with the top layer is lost because of incidents from exterior
Scalability	The system's architecture is scalable at every layer, with no obstacles in terms of devices to be monitored or controlled, and with no obstacles in terms of location
Flexibility in operation	Interaction with the system can be done over browser from any place, anytime, by anyone that has permission to do so
Efficiency	Commissioning is done faster and simpler, as well as the maintenance and servicing
Secure communication	All units in the architecture have embedded features for security, with encryption possibilities for sensitive data, and the architecture facilitates the implementation of blockchain and DLT technologies for enhanced security; this was one of the reasons to avoid inclusion of units with low power or memory, because those are weak links in the chain
Low cost	Cost is proportional to the size of the installation, but the cost per unit installed at the beginning and over the lifecycle are considerably reduced due to easy commissioning and reduction of integrators
Homogeneity	Homogeneity of control and communication is horizontally ensured, to the level of each layer, and vertically, at the interfaces of layers
Integrability	The system ensures the aggregation of all subsystems to fulfill the complete function of monitoring and control from a central unit
Serviceability	Capacity of intervention from remote locations is facilitated by the proposed solution and many aspects can be adjusted via software
Easy configuration	Relative to traditional SCADA systems, configuration is here done from software in a big proportion and uses simple actions to perform, including user-friendly interfaces
Redundancy for safety	This characteristic was explained in several locations of the paper, and it is ensured at every layer of the system, from field to cloud
Preventability	Cloud computing facilitates integration of data analytics in a natural way, as well as more sophisticated AI models, based on machine learning (e.g., classification algorithms [49]), or simpler neural networks (e.g., multi-layer perceptron is sufficient powerful for preventive maintenance), or deep neural networks (e.g., convolutional or recurrent neural networks); this subject is not treated in the paper because such problem requires access to historical data, but a good example for the system is the capacity to anticipate leakages of the pumps by measuring pressure, volume flow, and temperature and collect data from the cloud from all pumps in the network, where, with a 2-layer perceptron and real data, a high accuracy model can be trained to estimate when a certain pump will most likely leak
Architectural reliability	Reliability of the architecture was clearly demonstrated in the paper, highlighting this performance at each layer of the architecture; the proposed solution is based on proven industrial technologies
Necessary functionality	This characteristic is obvious, and there is nothing additional in conjunction with traditional SCADA systems, with the indication that there is no barrier to include any kind of sensor of actuator in the system for being monitored and controlled
Connectivity	The use of flow-based programming tools (e.g., node-RED) facilitate a lot of connectivity of various devices and services, in an easy way, directly to the cloud
Easy commissioning	Less integrators are necessary, and most of the tasks can be done from the office, with no need to locate high skilled staff in the field

Advantages with respect to the traditional SCADA systems have been already commented in Section 5.2, as well as in several other places in the paper, therefore they will not be repeated in this section of the paper.

7. Conclusions

Design and engineering of cloud computing and IoT-driven SCADA systems is a new trend in WWT installations for monitoring and control. Such novel architectures come with clear advantages and open new doors of opportunity for innovative business models in the WWT industry. Traditional and disruptive SCADA systems will co-exist until the disruptive technology will evolve to higher levels of reliability and cybersecurity. This might take a decade or more, but traditional technologies will be replaced by the newcomers. New businesses will emerge to provide user-friendly interfaces on web for SCADA applications, and this will accelerate the process of adoption of the new SCADA architectures.

A major contribution of our paper to the literature in the field is the introduction of a systematic methodology to design SCADA systems in the conceptualization phase under the condition of complex requirements. This kind of approach was not yet reported for this stage of design, where empiricism, experience, and trial-and-error experiments dominated the scene. The methodology can be extended to any other engineering system, with adaptation for the particular cases. However, as it was highlighted several times in the paper, this methodology does not cover quantitative optimization, which is a subject of parameter optimization. Usually, this task follows conceptual design, and is reflected during detailed design and prototype experimentation.

The conclusion is that such systems are complex, and they cannot be designed and developed from scratch by a single player. Therefore, the future is of open innovation and innovation ecosystems, of open platforms, where collective contributions will mainly count to the advancements in this field. Scientific research in this area will be necessary, but mainly focused on very narrow topics. In terms of architectural design, interfacing and configuration design, systematic processes for planning, analysis, and conflict solving will still play an important role, simply for the reason of not omitting essential aspects in the design due to the complexity of the integrated technologies in such systems. The lifecycle approach and thinking are desirable, despite the fact that they come with many uncertainties and with an n-dimensional space for optimization. Reduction of dimensionality is very much appreciated for system engineering, therefore working on a better understanding of various implications of the large set of objective-function is necessary. In some situations, one could introduce simpler strategies to reduce dimensionality. In order cases this goal is not possible; hence, innovations are required to overpass complexity. One of these innovations we had in mind is the analysis of motricity and dependency of the objective-functions, with the consideration of the 80-20 rule. It was not the case to activate it in this research, but it might work for other cases.

Future researches must be concentrated on solutions for reducing latency in data transfer for cloud computing, especially for critical WWT infrastructures, as well as for local CPSs that embed more memory and security capacities.

Author Contributions: Conceptualization, S.B.; methodology, S.B.; software, M.M. and M.P.; validation, M.M., G.V. and E.B.; formal analysis, S.B.; investigation, M.M.; resources, G.V.; data curation, E.B.; writing—original draft preparation, S.B. and M.M.; writing—review and editing, E.B.; visualization, M.M.; supervision, S.B. and G.V.; project administration, G.V.; funding acquisition, S.B. and G.V. All authors have read and agreed to the published version of the manuscript.

Funding: This research received no external funding.

Institutional Review Board Statement: Not applicable.

Informed Consent Statement: Not applicable.

Data Availability Statement: Not applicable.

Conflicts of Interest: The authors declare no conflict of interest.

References

1. Strzelecka, A.; Ulanicki, B.; Koop, S.; Koetsier, L.L.K.; Elelman, R. Integrating water, waste, energy, transport and ICT aspects into the smart city concept. In Proceedings of the 18th Conference on Water Distribution System Analysis, Cartagena, Colombia, 23–28 July 2016; Volume 186, pp. 609–616.
2. Rezai, A.; Keshavarzi, P.; Moravej, Z. Key management issue in SCADA networks: A review. *Int. J. Eng. Sci. Technol.* **2017**, *20*, 354–363. [CrossRef]
3. Stouffer, K.; Falco, J.; Kent, K. Guide to Supervisory Control and Data Acquisition (SCADA) and Industrial Control Systems Security. *NIST Spec. Publ.* **2006**, *27*, 800–882.

4. National Communications Systems. *National Detecting Integrity Attacks on SCADA Systems, Communications Systems*; Supervisory Control and Data Acquisition (SCADA) Systems: Virginia, VA, USA, 2005; pp. 1–76.

5. Lakhoua, M.N.; Jbira, M.K. Project management phases of a SCADA system for automation of electrical distribution networks. *Int. J. Comput. Sci. Issues* **2012**, *9*, 157–192.

6. United States Environmental Protection Agency. *Online Water Quality Monitoring in Distribution Systems for Water Quality Surveillance and Response Systems*; United States Environmental Protection Agency: Cincinnati, OH, USA, 2018; pp. 1–86.

7. Gray, A.D.L.; Pisica, I.; Taylor, G.A.; Whitehurst, L. A standardized modular approach for site SCADA applications within a water utility. *IEEE Access* **2017**, *5*, 17177–17187. [CrossRef]

8. Bangemann, T.; Karnouskos, S.; Camp, R.; Carlsson, O.; Riedl, M.; McLeod, S.; Harrison, R.; Colombo, A.W.; Stluka, P. State of the Art in Industrial Automation. In *Industrial Cloud-Based Cyber-Physical Systems*; Colombo, A., Ed.; Springer: Cham, Switzerland, 2014. [CrossRef]

9. Harjunkoski, I. Process Automation Systems. Available online: https://mycourses.aalto.fi/pluginfile.php/552398/mod_folder/intro/2017-09-29%20Process%20Automation%20actual%20IH.pdf (accessed on 3 December 2020).

10. Janicke, H.; Nicholson, A.; Webber, S.; Cau, A. Runtime-monitoring for industrial control systems. *Electronics* **2015**, *4*, 995–1017. [CrossRef]

11. Chen, Y.; Han, D. Water quality monitoring in smart city: A pilot project. *Autom. Constr.* **2018**, *89*, 307–316. [CrossRef]

12. Nguyen, V.H.; Tran, Q.T.; Besanger, Y. SCADA as a service approach for interoperability of micro-grid platforms. *Sustain. Energy Grids Netw.* **2016**, *8*, 26–36. [CrossRef]

13. Börjesson, R.; Eriksson, E.; Wangel, J. ICT practices in smart sustainable cities-In the intersection of technological solutions and practices of everyday life. In Proceedings of the EnviroInfo and ICT for Sustainability 2015, Copenhagen, Denmark, 7–9 September 2015; pp. 317–324. [CrossRef]

14. Kramers, A.; Höjer, M.; Lövehagen, N.; Wangel, J. Smart sustainable cities—exploring ICT solutions for reduced energy use in cities. *Environ. Model. Softw.* **2014**, *56*, 52–62. [CrossRef]

15. Kudva, S.; Ye, X. Smart cities, big data, and sustainability union. *Big Data Cogn. Comput.* **2017**, *1*, 4. [CrossRef]

16. Kharrazi, A.; Qin, H.; Zhang, Y. Urban big data and sustainable development goals: Challenges and opportunities. *Sustainability* **2016**, *8*, 1293. [CrossRef]

17. Bibri, S.E. The IoT for smart sustainable cities of the future: An analytical framework for sensor-based big data applications for environmental sustainability. *Sustain. Cities Soc.* **2018**, 230–253. [CrossRef]

18. Park, E.; del Pobil, A.P.; Kwon, S.J. The role of internet of things (IoT) in smart cities: Technology roadmap-oriented approaches. *Sustainability* **2018**, *10*, 1388. [CrossRef]

19. Koo, D.; Piratla, K.; Matthews, J.C. Towards Sustainable Water Supply: Schematic Development of Big Data Collection Using Internet of Things (IoT). In Proceedings of the International Conference on Sustainable Design, Engineering and Construction, Chicago, IL, USA, 13 May 2015; Volume 118, pp. 489–497.

20. Mounce, S.R.; Pedraza, C.; Jackson, T.; Linford, P.; Boxall, J.B. Cloud based machine learning approaches for leakage assessment and management in smart water networks. In Proceedings of the 3rd International Conference on Computing and Control for the Water Industry, Delhi, India, 29–30 July 2020; pp. 43–52.

21. O'Donovan, P.; Coakley, D.; Mink, J.; Curry, E.; Clifford, E. Waternomics: A cross-site data collection to support the development of a water information platform. In Proceedings of the 13th Computer Control for Water Industry Conference, Leicester, UK, 2–4 September 2015; pp. 458–463.

22. Combs, L. Cloud Computing for SCADA. Available online: https://www.controleng.com/articles/cloud-computing-for-scada/ (accessed on 3 December 2019).

23. Bristol Is Open Ltd. Bristol Is Open—Open Programmable City. Available online: https://www.bristolisopen.com (accessed on 1 December 2019).

24. Narayanan, L.K.; Sankaranarayanan, S. IoT enabled smart water distribution and underground pipe health monitoring architecture for smart cities. In Proceedings of the IEEE 5th International Conference for Convergence in Technology, Pune, India, 29–31 March 2019.

25. Roy, J.K.; Mukhopadhyay, S.C. Monitoring water in treatment and distribution system. In *Next Generation Sensors and Systems*; Mukhopadhyay, S.C., Ed.; Springer: Berlin/Heidelberg, Germany, 2016; Volume 16, pp. 257–287. [CrossRef]

26. Dugyala, R.; Reddy, N.H.; Kumar, S. Implementation of SCADA through cloud-based IoT devices—The initial design steps. In Proceedings of the 5th International Conference on Image Information Processing, Waknaghat, India, 15–17 November 2019; pp. 367–372.

27. Baker, T.; Asim, M.; MacDermott, A.; Iqbal, F.; Kamoun, F.; Alfandi, O.; Hammoudeh, M. A secure fog-based platform for SCADA-based IoT critical infrastructure. *Softw. Pract. Exp.* **2020**, *50*, 503–518. [CrossRef]

28. Altshuller, G. *The Innovation Algorithm*; Technical Innovation Center: Worcester MA, USA, 2000.

29. You, T. The Curse of Dimensionality. Available online: https://towardsdatascience.com/the-curse-of-dimensionality-50dc6e49aa1e (accessed on 10 September 2020).

30. TRIZ-CM, UTCN. Available online: http://193.226.17.76:8080/sts291-mvc/tool_cmx.do?aProject=1&aSet=1&aAct=1&aTarget=1&aActivityName=1 (accessed on 20 September 2020).

31. Brad, S.; Murar, M. Novel architecture of intelligent axes for fast integration into reconfigurable robot manipulators: A step towards sustainable manufacturing. *Rom. J. Tech. Sci. Appl. Mech.* **2013**, *58*, 107–128.
32. NPM. Node–Red-Contrib-s7. Available online: https://www.npmjs.com/package/node-red-contrib-s7 (accessed on 11 May 2020).
33. NPM. Node-Red-Contrib-Mqtt-Broker. Available online: https://www.npmjs.com/package/node-red-contrib-mqtt-broker (accessed on 3 December 2018).
34. IBM. IBM Cloud. Available online: https://www.ibm.com/cloud/ (accessed on 5 November 2018).
35. MQTT Organization. Available online: http://mqtt.org/ (accessed on 1 February 2021).
36. Renuka, D.; Pooja, J.; Shristhi, D.; Harshada, D.; Wagh, K.S. Secure distributed storage system for large-scale IoT data using blockchain. *J. Emerg. Technol Innov. Res.* **2019**, *6*, 597–601.
37. Rauchs, M.; Glidenn, A.; Gordon, B.; Pieters, G.; Recanatini, M.; Rostand, F.; Vagneur, K.; Zhang, Z.B. *Distributed Ledger Technology Systems: A Conceptual Framework*; Cambridge Center for Alternative Finance, University of Cambdrige: Cambdrige, UK, 2018.
38. Stanciu, A. Blockchain based distributed control system for edge computing. In Proceedings of the 2017 21st International Conference on Control Systems and Computer Science, Bucharest, Romania, 29–31 May 2017; pp. 667–671. [CrossRef]
39. Brad, S. Improving the use of AIDA method. *Appl. Math. Mech.* **2007**, *50*, 4.
40. Brad, S. Complex system design technique. *Int. J. Prod. Res.* **2008**, *46*, 5979–6008. [CrossRef]
41. Wang, H.; Xie, M.; Goh, T. A comparative study of the prioritization matrix method and the analytic hierarchy process technique in quality function deployment. *Total. Qual. Manag.* **1998**, *9*, 421–430. [CrossRef]
42. Agnihotri, A. Extending boundaries of blue ocean strategy. *J. Strateg. Mark.* **2016**, *24*, 519–528. [CrossRef]
43. Siemens, A.G. SIMATIC WinCC Options. Available online: https://w3.siemens.com/mcms/human-machine-interface/en/visualization-software/scada/wincc-options/pages/default.aspx (accessed on 10 November 2020).
44. Siemens, A.G. SIMATIC IOT2000. Available online: https://www.siemens.com/global/en/home/products/automation/pc-based/iot-gateways/iot2000.html (accessed on 21 June 2020).
45. Hive, M.Q. Reliable Data Movement for Connected Devices. Available online: https://www.hivemq.com/ (accessed on 18 April 2020).
46. Comparative Analysis of MQTT Technologies. Available online: https://waterstream.io/product/comparison/?gclid=EAIaIQobChMIv6-D19Dk7wIVCbbtCh3vGAbkEAAYASABEgIXf_D_BwE (accessed on 18 April 2020).
47. IBM Bluemix (IBM Cloud). Available online: https://www.ibm.com/cloud/blog/announcements/bluemix-is-now-ibm-cloud (accessed on 18 April 2020).
48. Han, Y.; Liu, S.; Geng, Z.; Gu, H.; Qu, Y. Energy analysis and resources optimization of complex chemical processes: Evidence based on novel DEA cross-model. *Energy* **2021**, *218*. [CrossRef]
49. Han, Y.; Qi, W.; Ding, N.; Geng, Z. Short-time wavelet entropy integrating improved LSTM for fault diagnosis of modular multilevel converter. *IEEE Trans. Cybern.* **2021**. [CrossRef] [PubMed]

sustainability

Article

The Financing of Wastewater Treatment and the Balance of Payments for Water Services: Evidence from Municipalities in the Region of Valencia

Marcos García-López *, Joaquín Melgarejo and Borja Montano

University Institute of Water and Environmental Sciences, University of Alicante, 03690 San Vicente del Raspeig, Spain; jmelgar@ua.es (J.M.); borja.montano@ua.es (B.M.)
* Correspondence: marcos.garcialopez@ua.es

Abstract: Pollution from wastewater discharges requires the treatment of all wastewater to maintain water bodies in good condition, as well as the possibility of reusing this water. Thus, wastewater treatment is an activity that has developed significantly in the Region of Valencia and has significant costs, including energy, which represents the main economic cost and an important environmental cost. In this way, efficiency and adequate financing of this activity are essential to minimise our environmental impact. However, the main funding tool currently does not allow us to address this issue, so we have a wastewater treatment with a high environmental cost in the form of greenhouse gas emissions. This tool is part of the revenues of water services, so it is not entirely independent, but it also seeks to prevent households from paying too high a total price. This leads to a situation where changes are needed to improve the financing of the different water services, as the financial resources obtained are insufficient and do not allow the current environmental problems to be solved. The analysis shows the importance of an appropriate tariff structure, as well as the need to include aspects such as water pollution and energy costs in the wastewater treatment tariff.

Keywords: water prices; water tariffs; sanitation taxes; wastewater treatment costs; energy costs; household budgets

Citation: García-López, M.; Melgarejo, J.; Montano, B. The Financing of Wastewater Treatment and the Balance of Payments for Water Services: Evidence from Municipalities in the Region of Valencia. *Sustainability* **2021**, *13*, 5874. https://doi.org/10.3390/su13115874

Academic Editor: Charikleia Prochaska

Received: 19 April 2021
Accepted: 20 May 2021
Published: 24 May 2021

Publisher's Note: MDPI stays neutral with regard to jurisdictional claims in published maps and institutional affiliations.

1. Introduction

Water resources are fundamental to all human activity, so their efficient management is essential to achieve the environmental sustainability of society. Unfortunately, water is a scarce resource in some parts of the world, and expectations for the future are for a reduction of annual rainfall, increased periods of resource scarcity and increased incidence of extreme weather events [1]. It is therefore essential to maximise efficiency in the management of an essential resource whose situation is expected to worsen. This is the situation in the Mediterranean area, a region where droughts or periods of scarcity, through their economic, social and environmental impacts, are a major problem [2,3]. Droughts, compared to other phenomena such as floods or storms, are more difficult to detect and quantify, which, together with the prospect of increasing global temperatures, brings with it the need to optimise our use of water resources [3].

In this context, one of the most important activities is wastewater treatment and reuse, which has developed significantly in the Region of Valencia (Spain) as a response to the scarcity of resources [4]. Purification is compulsory according to European regulations, as it is of great importance to reduce pollution from discharges in order to maintain water bodies in good condition [5]. Thus, there are minimum water quality criteria that must be met in order for the discharge to be considered adequate [6]. Moreover, pollution reduction is not the only contribution of this activity, as through improved treatment it is possible to reuse reclaimed water, thus generating additional water resources and relieving pressure on water bodies [7]. On the other hand, however, the activity has negative aspects, mainly derived from the environmental impact of the activity. This arises from various factors such as the construction of facilities or the use of chemical products, but the high energy consumption required to carry out the activity stands out [8,9]. This energy cost, in addition to the pollution generated, as it comes largely from fossil fuels [4], is the main economic cost of the activity, so maximising the energy efficiency of this activity has both economic and environmental benefits [10]. In any case, wastewater treatment partially avoids the deterioration of the quality of water bodies by reducing the presence of pollutants in discharged waters, so it is not an activity that should be abandoned. In this sense, it is expected that, in the future, if the status of these bodies of water continues to deteriorate, the costs of wastewater treatment will increase, as the pollution to be removed from the wastewater will be greater [11]. Finally, it should be borne in mind that the choice of treatment is linked to energy consumption and economic cost [12]. Moreover, these differences are not constant; the cost can vary significantly depending on the size of the installation [13], as well as many other aspects [14].

This activity, of course, requires sufficient funding to be developed, and, as the Water Framework Directive [5] indicates, it is the pollutant's responsibility to pay for correcting it. In this case, it is the water consumer who must pay for treatment, as it is his or her water consumption that causes the need for treatment. Thus, in the Region of Valencia, financing is obtained through a tax, called the sanitation tax, which is levied on domestic and industrial water consumption. Therefore, this tax is mainly responsible for the treatment aimed at reducing pollution from discharges, as the treatments aimed at regenerating wastewater for reuse are usually financed by the end user of the water. In this region, the levy is applied at the same time as local tariffs, so some water price effects can be attributed to it. Of course, the main one is the cost recovery of the activity, which is essential, since, between investment, operational and maintenance costs, the costs are significant and the good condition and functioning of the infrastructure must be pursued [15]. Full cost recovery is very difficult to achieve due to the high energy consumption, but even so, it is necessary to raise the amount needed to maintain the activity [16,17]. The next price effect is to reduce consumption, since water is a normal good, and if its price increases, its consumption should decrease [18]. However, we must never forget that there is a basic consumption that cannot be reduced by price increases [19]. In addition, it may be found that the tariff does not adequately convey the cost of the service or the scarcity of water, such that the relatively small amount of the water invoice would not change consumption. This issue has two aspects of interest: firstly, the difficulty of predicting consumer response, with its respective influence on public revenues, and secondly, the need for additional measures, such as awareness-raising campaigns, to change the behaviour of water users [20]. There are many other effects of water pricing policy, but the design of water pricing policy depends largely on the economic, environmental, social and political situation [18]. In order to face these characteristics, water prices have a certain capacity to adapt. In this sense, the water tariff can take into account some specific aspects such as scarcity at the time of consumption [21], the time of year of consumption [22] or the number of household members [23], although additional information is missing when designing the tariff [23]. Among these factors, the household structure is of particular importance, because if prices are set without taking into account the number of household members, household budgets may be affected differently on the basis of this characteristic [24]. In other words, the tariff should aim for fairness and not overly affect household finances [25,26], although it may be the case that the water invoice is very low for the household [26].

All these details are relevant when designing the price of a water service, the financing of which enables water policy to be improved [17]. Currently, the sanitation tax in the Region of Valencia does not have consumption brackets but has a fixed service fee and a variable fee that is applied equally regardless of how much has been consumed. The feature it includes is that, depending on the size of the municipality, both components of the tax vary, so that smaller municipalities pay a lower amount than larger ones. Given the current situation of high energy consumption, with its respective environmental cost, the aim of this work is to analyse, with recent data, the energy consumption of the wastewater treatment plants in the Region of Valencia, as well as the situation of the sanitation tax. In this way, the aim is to provide knowledge with a view to its modification, which would make it possible to reduce the considerable environmental impact of an activity that we cannot do without. However, this work analyses a specific tariff structure, so without additional information, it may not be useful for other regions with different tariffs and environmental, social and economic situations. For this reason, when analysing the information available for the Region of Valencia, the tariff structure of other places will also be assessed, thus establishing the usefulness of the article for obtaining water revenues in an efficient way in other regions. With this objective in mind, the next section will explain the data used and the methodology followed. Then, the results will be commented on and discussed, and finally, the conclusions obtained will be provided.

2. Materials and Methods

In order to meet the proposed objective, a series of data from various sources is used to address the issue in a more complete way than with only specific data. Moreover, these data have been worked on in a specific way to address the current problem.

2.1. Materials

In terms of data, several sources have been used to carry out this work (see Supplementary Materials). Firstly, basic information on the wastewater treatment plants in the Region of Valencia is available from the EPSAR (Entidad Pública de Saneamiento de Aguas Residuales) website. Specifically, the analysis uses the average energy consumption (kWh/m^3), the population served (population equivalent), the quantity of wastewater treated (m^3/day), the quantity of wastewater originally designed to be treated in the plant (capacity of treatment plant measured as m^3/day) and the removal efficiencies in terms of Suspended Solids (SS), Biological Oxygen Demand (BOD) and Chemical Oxygen Demand (COD). The quantity of wastewater originally designed to be treated in the plant is constant, but the rest of the variables are 2018-year data obtained for each plant individually. Finally, it should be noted that these data are the official operating results of the facilities in the Region of Valencia. However, we do not go into detail on the treatments used on each plant because there are big differences between them, the treatments are diverse and therefore it is very difficult to find a general pattern.

On the other hand, information is included from the Household Budget Survey, which is elaborated annually by the National Statistics Institute (INE for its Spanish acronym). Four different editions of the survey are used, controlling by year, with the aim of increasing the available sample. Specifically, information is available for the years 2016, 2017, 2018 and 2019. This survey works on Spanish households budgets, so it is not specific to water resources, but it allows to include data on the price paid for water ($EUR/year$ or EUR/m^3) or for sanitation services ($EUR/year$ or $€/m^3$), the number of household members, household income ($EUR/year$) and water consumption ($m^3/year$). In addition, due to the differences in sanitation charges depending on the size of the municipality, the latter variable is also be included in the analysis. In order to eliminate cases that are very far from the average, which would distort it, several criteria have been followed. Firstly, households with an annual income of more than EUR 180,000 or an invoice of more than 25% of income have been eliminated. Households with a unit price of more than EUR 6, an annual invoice of more than EUR 1000 or an annual invoice per person of more than EUR 400 have also been eliminated. Finally, in terms of consumption, households with an annual consumption of more than 1000 m^3 or with an annual consumption per person of more than 400 m^3 have been eliminated. Unfortunately, households are not linked to any particular city, as this is part of ensuring the anonymity of the survey respondents, so it is not possible to take into account the particular characteristics of each city such as the number of households in the city or supply water pricing system.

To these two main sources of information were added the tariff for the sanitation canon of the Region of Valencia (from the management report of EPSAR [4]) and the current tariff for water supply of the city of Alicante (from the website of Aguas de Alicante), which were a city of this region and allow us to observe differences in the way revenues are obtained. Information is also included on the cost recovery of the hydrographic confederations present in the Region of Valencia, such as those of the Júcar and Segura.

2.2. Methods

With the above-mentioned data, the analysis is carried out in two distinct parts. Firstly, the available data on the wastewater treatment plants will be used in order to show some aspects of interest. This will be done through maps created with the GeoDa program, thus taking advantage of the geographical location of the facilities in the region. The data are not represented on a map as such, but, due to the use of coordinates, the figures included have the same shape as the Region of Valencia. After this, the wastewater treatment plants are divided into different groups through a cluster analysis. Specifically, this analysis is carried out on the basis of all the variables from the previous section. However, in order not to give too much importance to the removal efficiencies, what is included is the average of the 3. Thus, the classification is done by 4 variables: the average energy consumption, the quantity of wastewater treated, the proportion of the project flow used and the average removal efficiency. The number of clusters has been set to 5, the case grouping method used is K-means and the measure of association used is the Euclidean distance.

Regarding the data on households, tariffs and cost recovery of the hydrographic confederations, the main tool will consist of commenting on basic data. Thus, a table will be shown with data on households from the Household Budget Survey. This table will contain the different variables of interest depending on the size of the municipality in order to make comparisons with the structure of the sanitation tax. As for the tariffs and cost recovery data, these will be shown in tables before the data per household to contextualise the situation.

3. Results

Based on the above data and methods, it is possible to pursue the proposed objective. Thus, this section shows the results obtained, which allow us to generate discussion and draw certain conclusions.

3.1. Energy Consumption of Wastewater Treatment Plants

The aspect of wastewater treatment that we would like to highlight is its high energy consumption, which, as mentioned above, is associated with significant economic costs and greenhouse gas emissions. In other words, improving the energy efficiency of this activity and replacing its energy consumption based on fossil fuels with consumption from renewable sources are key aspects. In this sense, the first point is to show the energy consumption of the wastewater treatment plants in the Region of Valencia, which is done by means of Figure 1. This map shows the flow treated per year (size of the circle) and the average energy consumption (colour of the circle). As can be seen, the larger plants have a low energy consumption, but these are not very numerous, as there is a large number of small installations. Thus, a large dispersion can be observed between plants of similar size, with the northern part of the region not having a particularly high average energy consumption despite the high number of small plants. In the central part, although there is a greater number of plants with high consumption, there is also a large number with low consumption. However, the situation in the southern part of the region is very different, as the proportion of plants with high consumption is higher than in the other two parts of this region. In other words, it is possible that there is some important difference that conditions the activity in this region.

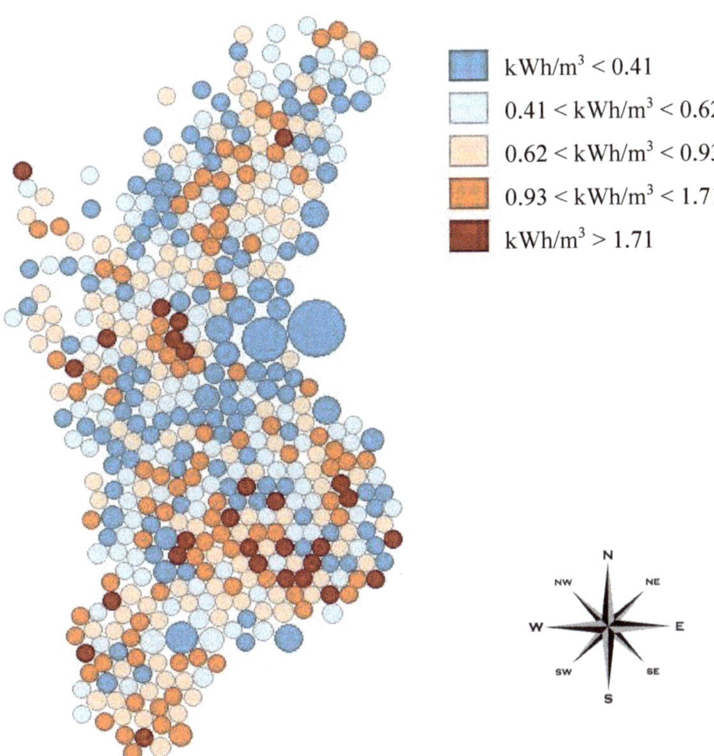

Figure 1. Map of wastewater treatment plants in the Region of Valencia, with the size of the circles indicating the amount of wastewater treated and the colour indicating energy consumption. Source: own elaboration based on EPSAR data.

For more information about the plants, Figure 2 shows the population served (size of the circle) and the proportion of the project flow utilised (colour of the circle). On this occasion, what we can see is how, in general, the installed capacity is not being utilised. In this sense, the utilisation of the installation cannot be forced, and it could be the case that certain plants are designed to be able to withstand the flow of wastewater at the time of peak consumption. For this reason, in annual terms, these plants may appear to be underutilised, as for a large part of the year the installation does not receive a large amount of wastewater. However, even in this case, there are a significant number of plants with very low-capacity utilisation, such that more treatment capacity is available than necessary. Moreover, this wasted capacity is concentrated in smaller plants, especially in the northern part of the region.

Finally, Figure 3, in order to provide information about the removal efficiencies, shows the amount of wastewater treated (size of the circle) and the average removal efficiencies (colour of the circle). Thus, we can see a clear relationship between size and yields, with larger plants having higher yields. In fact, among the small plants, there are some that do not meet the criteria set out in the regulation [6]. These are small installations that have a low impact associated with them, but even so, they have an impact that is not adequately addressed. In addition, most of the plants with low disposal yields are concentrated in the northern and central parts of the region. This fits with the previous result of higher energy consumption in the southern part of the region. In other words, the southern facilities have higher energy costs in exchange for improved disposal efficiencies and therefore less pollution from discharges and easier wastewater reuse.

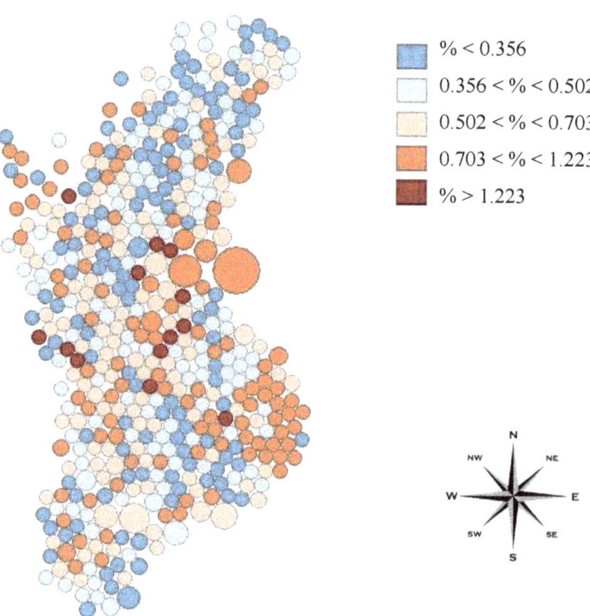

Figure 2. Map of wastewater treatment plants in the Region of Valencia, with the size of the circles indicating the population served and the colour indicating the percentage of installed capacity utilisation. Source: own elaboration based on EPSAR data.

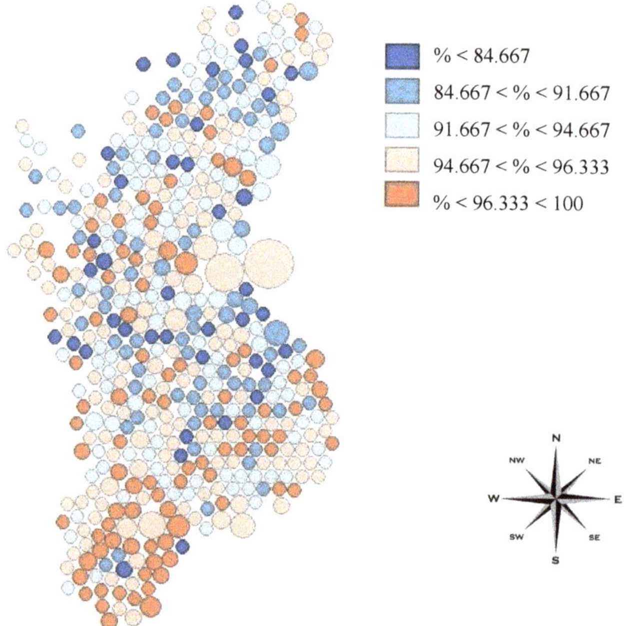

Figure 3. Map of wastewater treatment plants in the Region of Valencia, where the size of the circles indicates the amount of wastewater treated and the colour indicates the percentage of disposal performance (measured as the average of SS, BOD and COD). Source: own elaboration based on EPSAR data.

On the other hand, in order to expand the information about the wastewater treatment plants in the region, they have been grouped by means of a cluster analysis, whose results are shown in Table 1. The grouping and basic data of the five clusters created are in line with the information obtained from Figures 1–3, as the grouping is closely related to the size of the facilities. Thus, the first group contains 428 plants, representing 88.61% of the total, which, added to the 38 plants in the second group, makes a total of 96.48%. From here on, the groups are made up of plants of increasing size, until the last group is made up of one plant that is significantly larger than all the others. It can also be seen that the plants in the group have the lowest project flow utilisation, the worst disposal performance and the highest energy consumption of all. Of course, within this group, there is significant heterogeneity, which could be observed thanks to the maps. In any case, it is worth noting that there is a large number of small plants that are very difficult to operate efficiently, which has led in the past to proposals to concentrate flows as much as possible to reduce the use of small plants [9].

Table 1. Basic data on the groups of treatment plants in the Region of Valencia.

Variable/Cluster	1	2	3	4	5
Population served (Inhabitant equivalents)	1889.89	36,618.50	125,616.00	308,637.67	852,799.00
Proportion of project flow used (%)	0.55	0.58	0.63	0.73	1.06
Energy consumption (kWh/m^3/year)	0.80	0.58	0.43	0.28	0.23
Wastewater treated (m^3/year)	165,664.23	2,816,987.8	8,748,412.5	24,576,903	77,346,001
SS removal efficiency (%)	91.32	95.97	95.77	95	97
BOD removal efficiency (%)	94.64	96.53	97.46	97.67	98
COD removal efficiency (%)	90.01	92.97	93	93	94
Removal efficiencies (Mean SS, BOD and COD, %)	91.99	95.16	95.41	95.22	96.33
Number of plants	428.00	38.00	13.00	3.00	1.00
Percentage of total plants	88.61	7.87	2.69	0.62	0.21
Cumulative percentage	88.61	96.48	99.17	99.79	100.00

Source: own elaboration based on EPSAR data.

3.2. The Situation of Households and Water Supply and Treatment Tariffs

Given the current situation of the wastewater treatment plants in the Region of Valencia, where high energy consumption represents a significant economic and environmental cost, we now analyse the financial situation of wastewater treatment and reuse. Firstly, Table 2 [27,28] and Table 3 [29,30] come from the hydrological plans of the hydrographic confederations of the Júcar and Segura, both of which are present in the Region of Valencia, and show the cost recovery of both confederations. In general terms, cost recovery is low and limits the capacity to act. In terms of reuse, in one case, cost recovery is directly non-existent, while in the other case, there is some recovery, although it is quite low. For collection and treatment in public networks, the share of adequately financed costs is much higher, although there is still a significant lack of funding. Finally, in terms of total costs, we can see that an improvement in cost recovery is expected for the next hydrological cycle, but even with this improvement, the funds available are very limited, which greatly limits the capacity of the basin organisations to introduce modifications if they do not receive greater funding.

Table 2. Cost recovery for reuse, collection and treatment in public networks and total cost recovery of the Júcar Hydrographic Confederation for the hydrological cycles 2015–2021 and 2021–2027.

Cost Recovery		Financial Costs		Total Costs	
	Júcar Hydrographic Confederation	2015–2021	2021–2027	2015–2021	2021–2027
Reuse	Urban (garden irrigation)	-	-	-	-
	Agriculture/ranching	0%	0%	0%	0%
	Industry (golf)/energy	0%	0%	0%	0%
Collection and treatment in public networks	Urban supply	83%	-	75%	83.40% *
	Industry/energy	83%	-	75%	79.81% *
	Confederation Totals (Including all water services)	84%	93%	78%	87%

* These data include the economic and environmental costs, but not the resource cost. Source: CHJ, 2015, and CHJ, 2019.

Table 3. Cost recovery for reuse, collection and treatment in public networks and total cost recovery of the Segura Hydrographic Confederation for the hydrological cycles 2015–2021 and 2021–2027.

Cost Recovery		Financial Costs		Total Costs	
	Segura Hydrographic Confederation	**2015–2021**	**2021–2027**	**2015–2021**	**2021–2027**
Reuse	Urban (garden irrigation)	-	-	-	-
	Agriculture/ranching	5%	4%	3%	2%
	Industry (golf)/energy	53%	53%	53%	53%
Collection and treatment in public networks	Urban supply	80%	70%	46%	42%
	Industry/energy	82%	70%	47%	42%
	Confederation Totals (Including all water services)	83%	82%	57%	63%

Source: CHS, 2015, and CHS, 2020.

Therefore, the possibility to introduce modifications will be left to EPSAR, the public entity responsible for wastewater treatment and reuse and therefore also responsible for raising the necessary revenues. As mentioned above, the tax establishes a price according to the size of the municipality, as shown in Table 4 [4]. As can be seen, municipalities pay a significantly lower price compared to large municipalities (EUR 0.321 to EUR 0.441). This is related to the fact that the pollution load of wastewater from smaller municipalities is lower than in larger municipalities. However, considering that larger plants are more efficient in terms of energy consumption and disposal performance as well as economic costs, this is a striking fact. In addition, the economic characteristics of each municipality (or size of municipality) should also be considered in order to develop an efficient charge. These characteristics are assessed later, as we now compare the sanitation tax with one of the water supply tariffs. In particular, Table 5 [31] shows the tariff for the city of Alicante, the second-largest city in the Region of Valencia, which has a very particular design. Specifically, this progressive tariff presents an initial consumption bracket at a very low price so that the basic supply is guaranteed at a price that any household can afford. However, as household consumption increases, so does the price, reaching a maximum of EUR 2.85/m^3, which is a relatively high price that is, in fact, several times more than the sanitation charge applied to them (EUR 0.441). Although reaching this consumption bracket requires the consumption of a large amount of water, the previous brackets are also expensive at around EUR 2/m^3. On the other hand, this tariff has other special features, such as the fact that the service fee has discounts for the long-term unemployed or that the consumption brackets are modified according to the number of household members. Thus, the standard tariff shown in Table 5 only applies to normal households, i.e., not large families. In the case of large families, depending on the number of children, the brackets are modified so as not to overly penalise the consumption of such large households. Therefore, this tariff has a structure that allows for some customisation of the way revenues from the water supply service are obtained.

Table 4. Sanitation tax of the Region of Valencia.

	Sanitation Tax	
Population Brackets of Municipalities	**Consumption Quota (EUR/m^3)**	**Service Fee (€/year)**
500–3000	0.321	32.43
3001–10,000	0.376	39.75
10,001–50,000	0.412	43.81
More than 50,000	0.441	44.83

Source: EPSAR, 2020.

Table 5. Standard rate for domestic water consumption in the city of Alicante.

Household Customers *	Euros/m^3
From 0 to 12 m^3 per quarter	0.01
From 13 to 30 m^3 per quarter	0.7
From 31 to 45 m^3 per quarter	1.95
From 46 to 60 m^3 per quarter	2.02
From 61 m^3 per quarter onwards	2.85

* The service fee depends on the meter. There are some discounts, for example, in the fixed payment for being long-term unemployed or in the variable payment for being a large family (large family being understood as families with 3, 4, 5 or 6 or more members, with each of these having sections adapted to the number of members). Source: Aguas de Alicante, 2021.

Finally, as mentioned above, it is essential to know the characteristics of households in a region in order to properly assess a tariff or charge that will be applied to all of them. With this objective in mind, Table 6 shows the available data on households in the Region of Valencia by size of municipality. The values of the variables correspond to the average of all households. Firstly, if we talk about prices, we show the unit price per water tariff, per sanitation service and the total. The first of these prices does not show a clear pattern, but the payment for sanitation is higher in small municipalities. However, these payments present other sanitation services such as sewerage, so they do not accurately reflect the sanitation fee. It should be recalled that the fee is lower in smaller municipalities, so in reality, the fee payment is actually small compared to the total price. In this respect, the total price is between EUR 2.26 and EUR 2.80, of which only a maximum of EUR 0.441 (plus the service fee, which is independent of the amount) corresponds to the sanitation charge. If we take into account consumption per household and per person, we find that both consumptions are lower in the smaller municipalities, so that the higher unit price they present does not correspond to irresponsible consumption, at least in comparison with the rest of the households. If we also include household income in the analysis, we find that household income is high in small municipalities; in fact, income per person is the highest of all. Households in municipalities with between 50,000 and 100,000 inhabitants have a higher income per household, but their higher number of household members means that the income per person is lower than in smaller municipalities. Households of all other sizes have a lower income than the households discussed above. Finally, the key aspect of this analysis is the share of water and sanitation payments in total household income. Thus, since water supply is more expensive, it has a higher share of income than sanitation services. In total terms, this proportion is around 1.5% for households in municipalities with less than 50,000 inhabitants and around 1.3% for the rest. That is to say, if we value the price of water and sanitation payments together, we find that, in relative terms, households in small municipalities suffer a higher cost. However, this is not due to the sanitation charge, as this is lower in precisely these types of municipalities. In other words, in relative terms, smaller municipalities have a higher water supply price but a lower sanitation charge compared to larger municipalities. Therefore, as can be deduced, the situation is complex, and introducing modifications can be very complicated, but it is necessary given the current situation of scarce resources, high pollution and inefficiency in the wastewater treatment plants of the Valencian Community.

Table 6. Characteristics of households in the Region of Valencia by size of municipality.

Size of Municipality (Inhabitants) Variable	Less Than 10,000	Between 10,000 and 20,000	Between 20,000 and 50,000	Between 50,000 and 100,000	More Than 100,000
Unit price of the tariff (EUR/m^3)	1.78	1.68	1.44	1.45	1.64
Unit price of sanitation (EUR/m^3)	1.03	1.03	0.82	0.88	0.85
Total unit price (EUR/m^3)	2.80	2.71	2.26	2.33	2.49
Weight of the standard invoice on income (%)	0.97	0.96	1.02	0.84	0.88
Weight of sanitation on income (%)	0.53	0.53	0.55	0.47	0.44
Total weight on income (%)	1.50	1.49	1.57	1.31	1.32
Household consumption (m^3/year)	104.70	109.14	117.45	118.93	109.27
Consumption per person (m^3/year)	50.62	51.71	59.44	55.02	54.84
Household members	2.49	2.52	2.41	2.60	2.35
Household income (EUR/year)	25,409.04	23,006.56	21,697.31	26,787.04	22,371.14
Household income per person (EUR/year)	11,671.59	10,773.02	10,490.80	11,369.70	10,677.71

Source: own elaboration based on INE data.

4. Discussion

The results obtained show some aspects of interest that are very important for the efficient treatment and reuse of wastewater. We have been able to observe how the energy consumption of the wastewater treatment plants in the Region of Valencia is very high, mainly due to the large number of small plants. However, it has also been observed that plants in the southern part of the region have a higher energy consumption than those in the north, but in return, they have higher disposal yields, so the quality of the outgoing water is better. This has a huge effect on the economic cost of this activity, as the energy cost is its main economic cost [10]. For this reason, it is convenient to analyse the source of income of this activity, which has shown us that there is a large difference between the structure of the sanitation tax and the tariffs applied to water supply. In this way, we find that the sanitation charge represents a small part of the total payments for water services or sanitation, which means that it is a very small part of household budgets. In this sense, and given the results obtained on the economic situation of households in the Region of Valencia, it would be worth considering making changes to the sanitation charge.

However, while the changes are appropriate, we must take into account the current design of the revenue source and how it relates to the current economic situation in the region. This is a fundamental aspect, as the revenue structure of water services is key to the efficient management of water resources [32,33]. In this sense, we can find different types of tariffs, among which linear tariffs and increasing tariffs stand out [34,35]. For this reason, the analysis about the structure of water tariffs and the aspects to be taken into account in wastewater treatment payments is of value beyond the study area. In the case of the Valencia Region, the tariff is linear, but the unit amount paid depends on the size of the municipality. Thus, the sanitation fee is lower in smaller municipalities, where the energy and economic cost of the activity is higher and the household income per person is higher than elsewhere. This is one of the aspects to be taken into account, as the energy cost is one of the most important in this activity. However, these smaller municipalities face a higher price for water supply than the rest. In other words, the situation is very complex, as the tariff for water supply and sanitation services is set at the local level, while the sanitation fee is the same for the whole region. Therefore, this difference implies that both types of payment are set independently and may have an unequal impact on households depending on the location. This is of great relevance, as the regional government seeks to ensure that public prices affect households in a balanced way. Unfortunately, in the absence of control over local prices, the regional price, which was subsequently established, is conditioned by them as long as it remains the objective of not overburdening household budgets. Given this situation, a simple increase in the charge can have important consequences on this objective and on the financing of the activity. It should not be forgotten that the cost to the household of the sanitation charge is actually quite small, and its increase could be difficult for users to perceive. In this respect, the importance of public acceptance of the activity should not be forgotten, as users' willingness to pay will be reduced if they do not correctly perceive the benefits obtained [36,37]. In any case, the results obtained for payment for supply and for sanitation services have shown us the large differences that can arise when water design revenues are designed at different territorial scales. In this respect, the balance between the two types of payment can vary greatly between regions. This is not a major problem as long as payments are not particularly high or there is no need for additional funding. However, it is a problem to be solved when tariff imbalances lead to problems of equity or financing of the service.

Given the lack of control over local payments, a sanitation tax capable of efficiently financing the activity of wastewater treatment and reuse is essential. In fact, it is not only a necessity of the Region of Valencia, but improvements must be introduced throughout Spain in order to adapt the activity to European regulations and achieve cost recovery [38]. In this sense, it is very convenient to assess whether the structure of this type of payment, which simply includes the size of the municipality, is the most appropriate. As discussed above, this simple structure is easy to manage, but it is not suitable for efficient revenue collection. This simple structure does not allow taking into account some aspects that have been shown to be relevant in determining the functioning of the activity. On the one hand, we should appreciate the fact that the tariff is linear, as it does not provide any disincentive to the water consumer. We should not forget that this price, in the eyes of the consumer, is part of the water invoice and reacts to it in conjunction with the water supply tariff. Thus, the ability of water pricing to stimulate efficient resource consumption is being underutilised. Linear tariffs, although not as common today as increasing block tariffs, still have a significant presence worldwide, and this type of analysis becomes relevant to study their modifications [34]. On the other hand, the wastewater treatment activity has some particularities that need to be considered. Firstly, the presence of pollutants in the water significantly conditions the operation of a treatment plant, so that a tariff that does not include this issue would be neglecting an important aspect of a treatment plant's performance [39]. Secondly, the situation is similar with respect to the energy cost, which has a significant environmental and financial cost and depends to a large extent on the efficiency of the treatment plant but has no influence on the tariff. These two aspects are an important part of a tariff that efficiently finances wastewater treatment. However, global tariffs do not include these relevant aspects, but their use is similar to those for water supply and does not take into account the particularities of this water service [34,39]. Thus, the analysis of the situation in the study region has allowed us to assess both the tariff structure and the aspects that should be included in it, which are universal issues that can be used elsewhere. The example of the Region of Valencia is a clear illustration of the impact of these issues on the operation of the plants and, therefore, of their financial costs and the need for appropriate tariffs. The situation in this region has shown us how the tariff must be adapted to specific situations in order to function efficiently. Otherwise, within that region, we would be faced with economic inefficiency and inequality of payments, which would be added to the already existing regional differences in Spain and Europe [40–43].

The current Valencian tax has an unequal impact on different households and does not generate sufficient revenue to address the energy consumption problems of wastewater treatment plants, especially the large consumption of smaller plants and the high greenhouse gas emissions associated with larger plants. Therefore, one alternative could be to modify the structure of the sanitation charge. The positive side of this possibility is that the different local tariffs have already worked on these issues, although in most cases the structure of these tariffs is reduced to introducing consumption brackets with increasing prices. This could be an alternative for the sanitation charge so as to discourage excessive consumption, which, apart from being an irresponsible act in environmental terms, leads to increases in the amount of wastewater to be treated in the purification facilities, with a corresponding increase in the size of the plants [15]. Thus, such consumption represents an unnecessary extraction of water resources and leads to an increase in the economic and environmental costs of wastewater treatment plants. A progressive sanitation charge, which would be coupled with similar tariffs, would address this issue. However, in any case, it would be interesting to have updated and accurate information on the economic and energy costs of each plant, as well as the population they serve, in order to be able to design the income from this activity in an efficient way.

Therefore, the current structure of the sanitation charge is adequate to achieve a certain level of revenue, but it is not adequate to finance the full (economic and environmental) costs of the activity, nor to have an equitable impact on the different users of the service. This is the reason for this analysis, which has allowed us to observe how the particularities of wastewater treatment require specific tariffs to obtain revenue in an efficient manner. Thus, aspects such as the pollution present in the wastewater and the energy cost of the installation should be assessed. Otherwise, the tariffs would not adequately convey the message they are intended to convey. In the same way that the water supply tariff seeks to inform about the financial cost of the service and the scarcity of the resource, the wastewater treatment tariff must convey the message of the high financial and environmental cost of the activity and the importance of minimising discharge pollution. In any case, it is very important to analyse the specific situation in a region before making tariff changes. Payments for water services should not be an excessive item in household budgets, but the financing of these services should be ensured in an efficient way. This issue, while relevant, is highly case-dependent, as the socio-economic situation may vary from case to case. At present, this is not only a problem of the Region of Valencia, but there is still a large worldwide presence of linear tariffs, but above all, the current wastewater treatment tariffs do not include the aspects discussed in this article. However, it should be noted that the inclusion of aspects related to the operation of the activity would imply a significant management cost, so this is an aspect that needs to be studied in depth.

5. Conclusions

The aim of this work was to analyse the current situation of the energy cost, with its respective economic cost, of the wastewater treatment and reuse activity in the Region of Valencia, as well as to determine the adequacy of the current sanitation tax used as a financing tool. The result is valuable information on what aspects should be taken into account when designing such a tariff. Thus, it has been found that, while the size of the installation is a key determinant of the energy and economic cost of the treatments, there are other aspects that must be taken into account. The possibility to geographically locate the available wastewater treatment plants has shown us that in the southern region, a higher energy cost is incurred in order to achieve a higher quality of the resulting water. The situation in the north is quite different, as the energy cost of the activity is lower, but it can also be observed that the outgoing water from the plants in that part of the analysis region has a lower quality.

Therefore, the size of the plant is a key aspect to consider when designing the tariff, but no geographical discrimination can be included in the form of financing. In any case, the particularities of this activity require a financing instrument that addresses the specific situation. In this sense, the current charge only takes into account the size of the municipality when setting a price. This may make some sense, but in practice it does not fit the situation we live in, where the environmental cost of wastewater treatment is unnecessarily high due to the lack of energy efficiency. For this reason, it would be interesting to obtain new revenues, for which it is necessary to look at the situation of households, which are the ones that provide the main financing. In doing so, we find that there is an imbalance in the payments for water services, as households in small municipalities suffer a higher price for water supply but enjoy a lower sanitation fee. In total, these households have a relatively high cost for the sum of water supply and sanitation services. In other words, while they contribute less funding to wastewater treatment, they pay more for local services. This shows a significant imbalance that is not controllable by the regional government, as with the exception of the sanitation tax, because prices are determined at a local level. However, we can highlight the fact that there are significantly higher payments on the one hand and relatively lower payments on the other hand. In this way, it would be interesting to evaluate the structural modification of the sanitation tax, as well as to improve communications between the different public administrations. In particular, it could be considered that the sanitation charge should include some component of the water supply tariffs, which try to adapt more to the situation. In this way, it could also be interesting to include consumption brackets in the charge, as excessive consumption implies greater wastewater purification, with its respective economic and environmental costs.

The issues addressed in this article are not unique to the Region of Valencia, but the information presented allows useful conclusions to be drawn for other regions. It has become clear how large differences can exist between regions in terms of payments for water services, which directly affects the equity of the tariff, which is one of the objectives of any tariff. Of course, obtaining the necessary revenues for the development of water services is the main objective of the diverse tariffs, but this must be done in a fair and balanced way, without overly affecting household finances and respecting as much as possible the principle that the polluter pays. In any case, while respecting these issues, the revenues of each water utility must take into account the particularities of each of them. Thus, when we talk about wastewater treatment, we find a series of aspects that, if not included in the tariff that finances the service, limit its efficiency. Thus, aspects such as the presence of pollutants in the water or the high energy cost of this activity are key determinants of the financial cost of wastewater treatment and should be part of the corresponding tariff. However, it should be noted that the size of the facilities is a particularly important issue, so it should not be ignored either. Therefore, there are several key aspects in the development of this activity that we need to consider when obtaining the revenues. This occurs, moreover, in a context where tariffs tend to have simple structures to facilitate their management. For this reason, the way to include these aspects in the wastewater treatment tariff must be administratively feasible.

This work opens an interesting line of research, because, although it has revealed the possibility of introducing modifications to the main financing tool for wastewater treatment, it does not have the necessary information to carry out a more in-depth analysis. The main limitations in terms of the information available are the availability of annual data, which prevents the seasonal aspect of water consumption from being considered, as well as the scarce information on the financial cost of the wastewater treatment plants. Therefore, it would be possible to continue developing this possibility with the aim of improving the management of an activity whose efficiency is so important.

Supplementary Materials: The following are available online at https://www.mdpi.com/article/10.3390/su13115874/s1.

Author Contributions: M.G.-L. is the main author of this research with the supervision of B.M. and J.M. In particular, B.M. contributed with data analysis, while J.M. contributed to data collection and to structuring the article. All authors have read and agreed to the published version of the manuscript.

Funding: This research was financed by the Office of the Vice President of Research and Knowledge Transfer of the University of Alicante, Spain (Marcos García-López has a scholarship for The Training of University Teachers from the University of Alicante, UAFPU2019-16).

Institutional Review Board Statement: Not applicable.

Informed Consent Statement: Not applicable.

Data Availability Statement: The Household Budget Survey is available on the Spanish National Statistics Institute website, which can be consulted in English, but the design of the survey is only available in Spanish. The data about wastewater treatment and sanitation tax come from the EPSAR website, while the tariff of Alicante can be consulted on the webpage of Aguas de Alicante. Lastly, the data about costs recovery come from the websites of both hydrographic confederations mentioned in the text (available in references).

Acknowledgments: This work was supported by the Office of the Vice President of Research and Knowledge Transfer of the University of Alicante, Spain, the University Institute of Water and

Environmental Sciences of the University of Alicante, the Water Chair of the University of Alicante-Alicante Provincial Council and the Hábitat5U network of excellence. We would also like to thank the editor and reviewers involved in the publication process for their valuable comments to improve the article.

Conflicts of Interest: The authors do not have financial or personal interests that could have influenced the paper.

References

1. Navarro, T. Water reuse and desalination in Spain–challenges and opportunities. *J. Water Reuse Desalination* **2018**, *8*, 153–168. [CrossRef]
2. Barella-Ortiz, A.; Quintana-Seguí, P. Evaluation of drought representation and propagation in regional climate model simulations across Spain. *Hydrol. Earth Syst. Sci.* **2019**, *23*, 5111–5131. [CrossRef]
3. Spinoni, J.; Barbosa, P.; Bucchignani, E.; Cassano, J.; Cavazos, T.; Christensen, J.H.; Christensen, O.B.; Coppola, E.; Evans, J.; Geyer, B.; et al. Future Global Meteorological Drought Hot Spots: A Study Based on CORDEX Data. *J. Clim.* **2020**, *33*, 3635–3661. [CrossRef]
4. EPSAR. "Memoria de Gestión". Ejercicio 2019. Entidad Pública de Saneamiento de Aguas Residuales de la Comunidad Valenciana, Valencia. 2020. Available online: https://www.epsar.gva.es/sites/default/files/2020-07/INFORME-DE-GESTION_2019.pdf (accessed on 28 March 2021).
5. EU. Directive 2000/60/EC of the European Parliament and of the Council Establishing a Framework for the Community Action in the Field of Water Policy 2000. Available online: https://eur-lex.europa.eu/legal-content/EN/TXT/?uri=CELEX:32000L0060 (accessed on 28 March 2021).
6. EEC. Directive 91/271/EEC of the Council Directive of 21 May 1991 Concerning Urban Waste-Water Treatment 1991. Available online: https://eur-lex.europa.eu/legal-content/ES/TXT/?uri=LEGISSUM%3Al28008 (accessed on 28 March 2021).
7. Melgarejo, J. Efectos Ambientales y Económicos de la Reutilización del Agua en España. *CLM Econ.* **2009**, *15*, 245–270.
8. Hernández-Sancho, F.; Molinos-Senante, M.; Sala-Garrido, R. Eficiencia energética, una medida para reducir los costes de operación en las estaciones depuradoras de aguas residuales. *Tecnol. del agua* **2011**, *31*, 46–54.
9. Albadalejo-Ruiz, A.; Martínez-Muro, J.L.; Santos-Asensi, J.M. Parametrización del consumo ener-gético en las depuradoras de aguas residuales urbanas de la Comunidad Valenciana. *TecnoAqua Lucca Italy* **2015**. Available online: https://www.tecnoaqua.es/media/uploads/noticias/documentos/articulo-tecnico-parametrizacion-consumo-energetico-depuradoras-agua-residuales-urbanas-comunidad-valenciana-tecnoaqua-es.pdf (accessed on 5 May 2021).
10. Albadalejo-Ruiz, A.; Trapote, A. Influencia de las Tarifas Eléctricas en los Costes de Operación y Mantenimiento de las Depuradoras de Aguas Residuales. *TecnoAqua* **2013**, *3*, 48–54.
11. McDonald, R.I.; Weber, K.F.; Padowski, J.; Boucher, T.; Shemie, D. Estimating watershed degradation over the last century and its impact on water-treatment costs for the world's large cities. *Proc. Natl. Acad. Sci. USA* **2016**, *113*, 9117–9122. [CrossRef]
12. Melgarejo, J.; Prats, D.; Molina, A.; Trapote, A. A case study of urban wastewater reclamation in Spain: Comparison of water quality produced by using alternative processes and related costs. *J. Water Reuse Desalination* **2016**, *6*, 72–81. [CrossRef]
13. Rowan, P.P.; Jenkins, K.L.; Howells, D.H. Estimating sewage treatment plant operation and maintenance costs. *J. (Water Pollut. Control Fed.)* **1961**, *33*, 111–121.
14. Marzouk, M.; Elkadi, M. Estimating water treatment plants costs using factor analysis and artificial neural networks. *J. Clean. Prod.* **2016**, *112*, 4540–4549. [CrossRef]
15. Renzetti, S. Municipal Water Supply and Sewage Treatment: Costs, Prices, and Distortions. *Can. J. Econ.* **1999**, *32*, 688. [CrossRef]
16. Tardieu, H.; Préfol, B. Full cost or "sustainability cost" pricing in irrigated agriculture. Charging for water can be effective, but is it sufficient? *Irrig. Drain.* **2002**, *51*, 97–107. [CrossRef]
17. Zetland, D.; Gasson, C. A global survey of urban water tariffs: Are they sustainable, efficient and fair? *Int. J. Water Resour. Dev.* **2013**, *29*, 327–342. [CrossRef]
18. Rogers, P. Water is an economic good: How to use prices to promote equity, efficiency, and sustainability. *Hydrol. Res.* **2002**, *4*, 1–17. [CrossRef]
19. Nauges, C. Is all domestic water consumption sensitive to price control? *Appl. Econ.* **2004**, *36*, 1697–1703. [CrossRef]
20. Zikos, D. Urban water dilemmas under the multi-dimensional prism of sustainability. *Trans. Bus. Econ.* **2008**, *8*, 413–422.
21. Hughes, N.; Hafi, A.; Goesch, T. Urban water management: Optimal price and investment policy under climate variability. *Aust. J. Agric. Resour. Econ.* **2009**, *53*, 175–192. [CrossRef]
22. Renzetti, S. Evaluating the welfare effects of reforming municipal water prices. *J. Environ. Econ. Manag.* **1992**, *22*, 147–163. [CrossRef]
23. Arbués, F.; Villanúa, I.; Barberán, R. Household size and residential water demand: An empirical approach. *Aust. J. Agric. Resour. Econ.* **2010**, *54*, 61–80. [CrossRef]
24. Dahan, M.; Nisan, U. Unintended consequences of increasing block tariffs pricing policy in urban water. *Water Resour. Res.* **2007**, *43*, 3. [CrossRef]
25. Rey, D.; Pérez-Blanco, C.D.; Escriva-Bou, A.; Girard, C.; Veldkamp, T.I.E. Role of economic instruments in water allocation reform: Lessons from Europe. *Int. J. Water Resour. Dev.* **2019**, *35*, 206–239. [CrossRef]

26. Arbués, F.; Barberán, R.; Villanúa, I. Price impact on urban residential water demand: A dynamic panel data approach. *Water Resour. Res.* **2004**, *40*, 11. [CrossRef]
27. Confederación Hidrográfica del Júcar, "Memoria". Plan Hidrológico de Cuenca 2015–2021. 2015. Available online: https://www.chj.es/es-es/medioambiente/planificacionhidrologica/Paginas/PHC-2015-2021-Plan-Hidrologico-cuenca.aspx (accessed on 29 March 2021).
28. Confederación Hidrográfica del Júcar, "Documentos iniciales. Memoria". Plan Hidrológico de la Demarcación Hidrográfica del Júcar. Revisión de Tercer Ciclo (2021–2027). 2019. Available online: https://www.chj.es/es-es/medioambiente/planificacionhidrologica/Paginas/PHC-2021-2027-Indice.aspx (accessed on 29 March 2021).
29. Confederación Hidrográfica del Segura, "Memoria". Plan Hidrológico de la Demarcación Hi-Drográfica del Segura 2015–2021. 2015. Available online: https://www.chsegura.es/es/cuenca/planificacion/planificacion-2015-2021/plan-hidrologico-2015-2021/ (accessed on 29 March 2021).
30. Confederación Hidrográfica del Segura, "Documentos iniciales. Memoria". Plan Hidrológico de la Demarcación Hidrográfica del Júcar. Revisión de Tercer Ciclo (2021–2027). 2020. Available online: https://www.chsegura.es/es/cuenca/planificacion/planificacion-2021-2027/el-proceso-de-elaboracion/ (accessed on 29 March 2021).
31. Aguas de Alicante, "Modificación de las Tarifas de Abastecimiento de Agua, Conservación de Con-Tadores, Contratación y Reposición en Alicante". Boletín Oficial de la Provincia de Alicante. 2021. Available online: https://www.aguasdealicante.es/documents/231378/2242840/BOP-TARIFAS-ALICANTE-24-03-2021.pdf/e5273503-86e7-80b8-d165-1c88068bb1cd (accessed on 29 March 2021).
32. Marques, R.C.; Miranda, J. Sustainable tariffs for water and wastewater services. *Util. Policy* **2020**, *64*, 101054. [CrossRef]
33. Molinos-Senante, M.; Hernandez-Sancho, F.; Sala-Garrido, R. Tariffs and Cost Recovery in Water Reuse. *Water Resour. Manag.* **2012**, *27*, 1797–1808. [CrossRef]
34. Damkjaer, S. Drivers of change in urban water and wastewater tariffs. *H2Open J.* **2020**, *3*, 355–372. [CrossRef]
35. Nauges, C.; Whittington, D.; El-Alfy, M. A simulation model for understanding the consequences of alternative water and wastewater tariff structures: A case study of fayoum, Egypt. In *Freshwater Governance for the 21st Century*; Springer Science and Business Media: Berlin, Germany, 2015; pp. 359–381.
36. Smith, H.M.; Walker, G. The Political Economy of Wastewater in Europe. In *Water Science, Policy, and Management: A Global Challenge*, 1st ed.; Dadson, S.J., Garrick, D.E., Eds.; John Wiley & Sons: Hoboken, NJ, USA, 2019; Volume 12, pp. 215–232. [CrossRef]
37. Simachaya, W. Wastewater tariffs in Thailand. *Ocean Coast. Manag.* **2009**, *52*, 378–382. [CrossRef]
38. Jodar-Abellan, A.; López-Ortiz, M.I.; Melgarejo-Moreno, J. Wastewater Treatment and Water Reuse in Spain. Current Situation and Perspectives. *Water* **2019**, *11*, 1551. [CrossRef]
39. Ruiz-Rosa, I.; García-Rodríguez, F.J.; Antonova, N. Developing a methodology to recover the cost of wastewater reuse: A proposal based on the polluter pays principle. *Util. Policy* **2020**, *65*, 101067. [CrossRef]
40. Valero, L.G.; Pajares, E.M.; Sánchez, I.M.R.; Pérez, J.A.S. Analysis of Environmental Taxes to Finance Wastewater Treatment in Spain: An Opportunity for Regeneration? *Water* **2018**, *10*, 226. [CrossRef]
41. Pajares, E.M.; Valero, L.G.; Sánchez, I.M.R. Cost of Urban Wastewater Treatment and Ecotaxes: Evidence from Municipalities in Southern Europe. *Water* **2019**, *11*, 423. [CrossRef]
42. Valero, L.G.; Pajares, E.M.; Sánchez, I.M.R. The Tax Burden on Wastewater and the Protection of Water Ecosystems in EU Countries. *Sustainability* **2018**, *10*, 212. [CrossRef]
43. García-Valiñas, M.; Arbués, F. Wastewater Tariffs in Spain. In *Oxford Research Encyclopedia of Global Public Health*; Oxford University Press: Oxford, UK, 2021.

sustainability

MDPI

Article

Estimation of Energy Recovery Potential from Primary Residues of Four Municipal Wastewater Treatment Plants

Eleni P. Tsiakiri [1], Aikaterini Mpougali [1], Ioannis Lemonidis [1], Christos A. Tzenos [2], Sotirios D. Kalamaras [2], Thomas A. Kotsopoulos [2] and Petros Samaras [1,*]

[1] Department of Food Science and Technology, School of Geotechnical Sciences, International Hellenic University, GR-57400 Thessaloniki, Greece; elenitsiak.85@hotmail.com (E.P.T.); aikmpougali@gmail.com (A.M.); lemon.dwrikos@yahoo.gr (I.L.)

[2] Department of Hydraulics, Soil Science and Agricultural Engineering, School of Agriculture, Aristotle University of Thessaloniki, GR-54124 Thessaloniki, Greece; catzenos@agro.auth.gr (C.A.T.); sotiriskalamaras@gmail.com (S.D.K.); mkotsop@agro.auth.gr (T.A.K.)

* Correspondence: samaras@ihu.gr

Abstract: Wastewater treatment plants have been traditionally developed for the aerobic degradation of effluent organic matter, and are associated with high energy consumption. The adoption of sustainable development targets favors the utilization of every available energy source, and the current work aims at the identification of biomethane potential from non-conventional sources derived from municipal wastewater treatment processes. Byproducts derived from the primary treatment process stage were collected from four sewage treatment plants in Greece with great variation in design capacity and servicing areas with wide human activities, affecting the quality of the influents and the corresponding primary wastes. The samples were characterized for the determination of their solids and fats content, as well as the concentration of leached organic matter and nutrients, and were subjected to anaerobic digestion treatment for the measurement of their biomethane production potential according to standardized procedures. All samples exhibited potential for biogas utilization, with screenings collected from a treatment plant receiving wastewater from an area with combined rural and agro-industrial activities presenting the highest potential. Nevertheless, these samples had a methanogens doubling time of around 1.3 days, while screenings from a high-capacity unit proved to have a methanogens doubling time of less than 1 day. On the other hand, floatings from grit chambers presented the smallest potential for energy utilization. Nevertheless, these wastes can be utilized for energy production, potentially in secondary sludge co-digestion units, converting a treatment plant from an energy demanding to a zero energy or even a power production process.

Keywords: screenings; fats; biogas potential; wastewater treatment plant; energy utilization; anaerobic digestion

Citation: Tsiakiri, E.P.; Mpougali, A.; Lemonidis, I.; Tzenos, C.A.; Kalamaras, S.D.; Kotsopoulos, T.A.; Samaras, P. Estimation of Energy Recovery Potential from Primary Residues of Four Municipal Wastewater Treatment Plants. *Sustainability* 2021, 13, 7198. https://doi.org/10.3390/su13137198

Academic Editor: Farooq Sher

Received: 1 June 2021
Accepted: 23 June 2021
Published: 27 June 2021

Publisher's Note: MDPI stays neutral with regard to jurisdictional claims in published maps and institutional affiliations.

1. Introduction

The activated sludge process has been identified as an efficient method for the treatment of a wide range of wastewaters for a long time. The process is based on the aerobic degradation of organics, while pretreatment is required for the upstream removal of suspended solids contained in raw wastewaters. The design of a wastewater treatment plant is mainly carried out in order to satisfy the corresponding effluent quality regulations while less efforts are taken towards energy requirements. Effluent quality is achieved at the cost of significant energy consumption. Aerobic processes are highly energy-demanding techniques with an average energy input reported to range from 0.30 to 1.89 kWh/m^3 depending on the treatment method used and the influent properties [1]. Secondary aeration accounts for about 50% of the total electricity demand and 25–40% of the plant operating costs [1–3]. Wastewater treatment plants are considered as the highest municipal energy consumers [4]. This issue is becoming crucial since the overall energy demand by wastewater treatment plants in European countries is estimated at 27 TWh/yr [5].

The increase of electricity costs in combination with the awareness of climate change has given rise to efforts aiming at the development of energy conservation policies and the implementation of energy-efficient measures that can contribute to the international sustainability development goals. Anaerobic digestion of excess sewage sludge represents a common practice in activated sludge processes, developed rather as a method of stabilization of wasted sludge than focusing on its energy valorization. Dedicated methodologies have been applied including energy surveillance, benchmarking, and auditing tools adapted to wastewater treatment plants, resulting in 20–40% savings in energy consumption [6]. However, these methods ignore the inherent energy content of raw wastewaters.

The energy content of raw wastewater has been measured by bomb calorimetry at about 14.7 kJ/g COD, exceeding almost nine times the electricity demand for its aerobic treatment [7]. Taking into account a daily production of about 100 g COD/person, the energy content of wastewaters for a city with 1 million inhabitants is estimated at around 1470 MJ/d; these estimations are carried out without considering the energy content of industrial wastewaters, while recent calculations report an energy content chemically bound to influents exceeding 153 KWh/person equivalent-year [8,9]. Nevertheless, the benefits of energy utilization of the contents of raw influents have not been given much attention so far, due to the traditional concept of wastewater considered as a waste requiring energy for its safe disposal and not as a valuable resource. Energy harvesting from raw influents might support efforts towards self-energy efficiency of wastewater treatment plants, shifting them from energy consumers to zero-energy plants or even to power producers.

Traditionally, anaerobic digestion of excess sludge is the primary energy harvesting method often applied in wastewater treatment plants. However, an alternative potential energy source is represented by the primary screenings including papers, wood, plastics, stones, metal pieces, rag, etc., and byproducts such as fats, oils, and grease (FOG). These materials can be directly valorized since they are separated during the primary treatment stages in existing units, without the need for the installation of additional facilities. Screenings are currently disposed in sanitary landfills, representing an additional operation cost while contributing to GHG emissions during their biological degradation in landfills [10].

The energy utilization of these byproducts has received less interest mainly due to their small amount compared to the much larger amounts of excess sludge [11]. For example, about 1.5 kg/person-equivalent of screenings have been estimated for 2007 in Germany [12], while in another study in France, average production was estimated to range from 0.53 to 3.49 kg/person [11]. However, the amount of screenings and FOGs is expected to increase due to intense urbanization, the enforcement of more strict EU regulations related to primary treatment yields, and the application of measures to alleviate the pressure due to increased influent loadings bringing a wastewater treatment plant close to the design capacity [13]. Moreover, the utilization of every available energy resource has been recently re-examined for various reasons. These include the development of advanced technological innovations for fine screening (enhancing screening removal), which is a pre-requirement in advanced membrane bioreactor units, as well as the adoption of sustainability principles by EU countries, which are associated with less waste diverted to landfills and the utilization of non-conventional wastes for energy and resource recovery [14,15].

Anaerobic digestion represents a challenging method for the energy valorization of non-conventional wastes from wastewater treatment plants, which is favored since it is a process conventionally applied for excess sludge treatment. However, very few studies have been reported in the literature dealing with the examination of biogas production potential of screenings and FOGs, mainly due to the non-homogeneous character of these wastes and the utilization of FOGs for biodiesel production [11,16]. The concentration of volatiles solids in screenings may exceed 90% of total solids [13,17], resulting in biogas production as high as 0.62 L/kg vs. [18]. The application of fine screenings in a municipal wastewater treatment plant and the following anaerobic digestion demonstrated a 40%

reduction of energy demand in the Netherlands [19]. On the other hand, the addition of FOGs collected from restaurants in a municipal sludge digester resulted in a 30 to 80% increase in biogas production [20].

Nevertheless, the biomethane production rate is greatly affected by the composition of screenings and FOGs. The characteristics of these wastes are related to a great number of factors such as the presence of a combined or separated sewer system, the number and the type of the sewer system pumping stations upstream to the treatment plant, the catchment area, and the primary stage installations. In addition, the properties of the wastes can be varied due to specific climatic conditions, consumers behavior, and time frameworks, including seasonal, weekly, or daily dimensions [11].

The objectives of this work include the examination of energy valorization of non-conventional wastes produced from municipal wastewater treatment plants, the determination of the corresponding biomethane potential during anaerobic digestion, and the identification of the relation to their composition and the type of primary treatment stage configuration. The primary target of the work is the justification of the beneficial role of screenings and FOGs towards energy utilization, as an alternative to currently used management methods of these side streams.

2. Materials and Methods

2.1. Samples Collection

Samples were collected from different areas of the primary treatment stage of four wastewater treatment plants (WWTP) located in Northern Greece, with various design capacities and servicing areas including a wide range of human activities. Samples identification and information on their origin are provided in Table 1.

Table 1. Identification of samples collected from the primary treatment stage of four wastewater treatment plants and their characteristics.

No	Wastewater Treatment Plant Identification		Substrate Origin	Symbol
	Design Capacity, m³/Day	Area		
1	8400	Urban, semi-rural, piscatorial	Screening from mechanical bar screen, 6 mm spacing	ASC
			Floatings from aerated grit chamber	AFL
2	11,400	Urban, rural	Screening from mechanical bar screen, 3 mm spacing	KSC
			Floatings from aerated grit chamber	KFL
3	23,400	Urban, rural, agro-industrial	Screening from mechanical bar screen, 50 mm spacing	LSCC
			Screening from mechanical bar screen, 15 mm spacing	LSCF
			Screenings from mechanical bar screen, 50 mm spacing used in septic tanks line	LSCST
			Floatings from aerated grit chamber	LFL
4	155,150	Urban, industrial	Screenings from mechanical bar screen, 10 mm spacing	TSC
			Floatings from aerated grit chamber	TFL
			Screenings from 5 mm bar screen of sludge from primary sedimentation	TPS
			Sludge anaerobic digester used for inoculation	I

About 1 kg of each sample was collected, placed in plastic bottles, and transferred to the laboratory for further analysis. Samples were stored at 4 °C in a constant temperature refrigerator until the time of analysis. Prior to storage, samples were subjected to milling using a knife mill in order to receive particles of similar size of about 0.1 mm.

2.2. Chemical Analysis

The analysis of the samples took place for the determination of total and volatile solids, and fats using standard methods, i.e., drying at 105 °C for total solids; thermal treatment at 550 °C for volatile solids; and Soxhlet extraction for total fats [21]. Water-leached organic matter and nutrients concentrations were determined by the addition of 0.3 g of dry solids in 20 mL of deionized water and stirring for 20 min, followed by centrifugation and determination of COD, N-NH$_4$, and P-PO$_4$ in the aqueous phase. Parameters were measured using a HACH-Dr Lange DR3900 spectrophotometer and the corresponding standard cuvette test kits, i.e., the LCK714 COD kit, the LCK303 ammonium kit, and the LCK049 orthophosphates test kit, respectively.

2.3. Biochemical Methane Potential

The biochemical methane potential (BMP) was determined by anaerobic digestion treatment performed in batch reactors. Glass reactors with a total volume of 322 mL and a working volume of 120 mL were used. In each reactor an amount of substrate containing 1 g of vs. was added followed by the inoculum to reach a working volume of 120 mL and achieve 202 mL headspace. Inoculum was collected from the anaerobic digester of the municipal wastewater treatment plant No. 4, with the highest design capacity operating for a long period. The glass reactors were sealed using rubber stoppers and aluminum caps suitable for retrieving gas samples. In addition, a blank sample was prepared, containing 120 mL of pure inoculum. The treatment for each of the 12 substrates along with the blank were carried out in triplicates. All reactors were initially flushed with pure nitrogen gas (N$_2$), in order to ensure anaerobic conditions. The batch reactors were incubated at a temperature of 37 $\pm$ 1 °C for 80 days.

The production of methane (CH$_4$) was determined by injecting gas samples from the reactors into a gas chromatographer (GC-2010plusAT, SHIMADZU, Kyoto, Japan) equipped with an appropriate detector and columns [22]. For each sample, 150 µL of gas were acquired from the headspace of the reactor with a gas-tight syringe outfitted with a pressure lock. During the first 10 days the reactors gas composition was monitored daily and after that on a bi-daily schedule. A standard gas mixture (60% CH$_4$, 40% CO$_2$) was utilized to determine the % concentration of CH$_4$ of each gas sample. The obtained peak area was compared to that of a standard gas mixture (60% CH$_4$, 40% CO$_2$) injected at atmospheric pressure in the chromatographer. The calculation of the volume of the produced CH$_4$ was carried out by the multiplication of the headspace volume of each reactor with the % concentration of CH$_4$ of each gas sample [23].

For each substrate, the maximum specific growth rate of the methanogens (μ_{max}) was derived from a graph of the natural logarithm of methane (CH$_4$) production as a function of time, and calculation of the maximum value from the tangent's angle was conducted [24]. The methanogens' maximum doubling time (T$_{doublemax}$) was calculated as the quotient of the natural logarithm of 2 (ln2) divided by the methanogens μ_{max} [25].

3. Results

The content of primary samples in total and volatile solids and fats is shown in Figure 1. Efforts were taken to collect all samples at the same time period, in order to exclude potential seasonal variations. As shown, samples presented high total solids content, while volatile solids ranged from as low as 20% up to 90% of the total solids.

In addition to solid and fat content, a significant role in the energy valorization of samples through anaerobic fermentation is played by the composition of leachable organic matter, as well as the concentration of nutrients, i.e., nitrogen and phosphorous; these data are provided in Figure 2 for the various samples.

Biomethane production potential (BMP) represents a crucial parameter for the assessment of energy valorization of samples by anaerobic fermentation. BMP is illustrated in Figure 3 in mL of methane produced per g of vs. for the various samples as a function of fermentation time, while total net methane production in L/kg vs. is given in Table 2, excluding the corresponding methane produced by the inoculum, i.e., the blank sample.

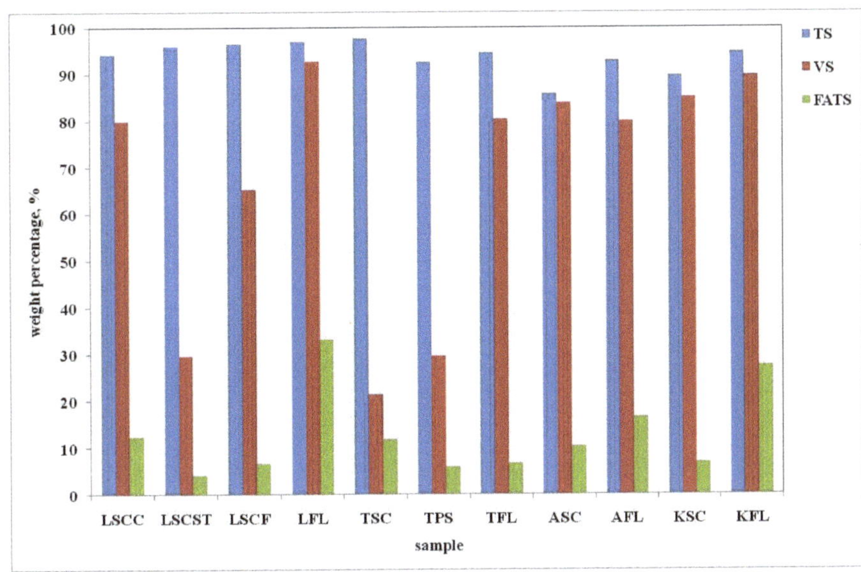

Figure 1. Total solids, volatile solids, and fats content in primary treatment samples from four WWTPs.

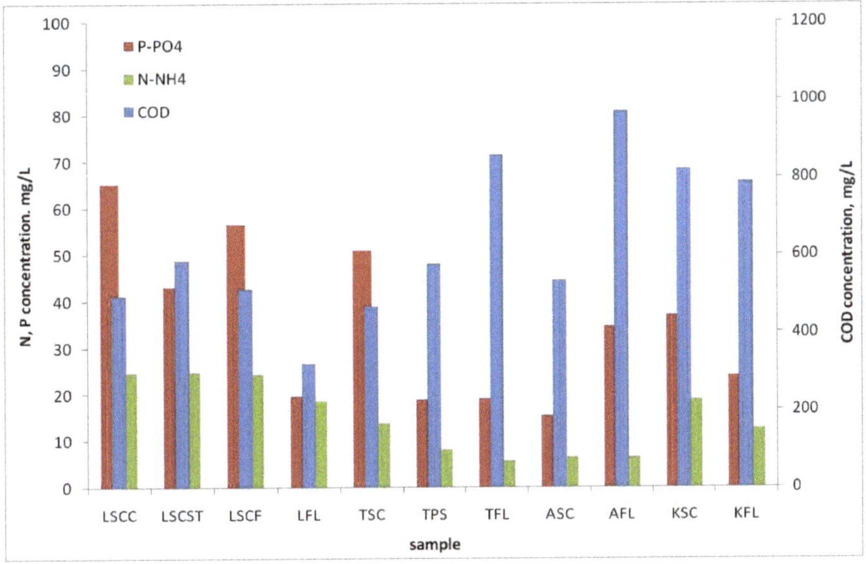

Figure 2. Leached COD, nitrogen, and phosphorous content in primary treatment samples from four WWTPs.

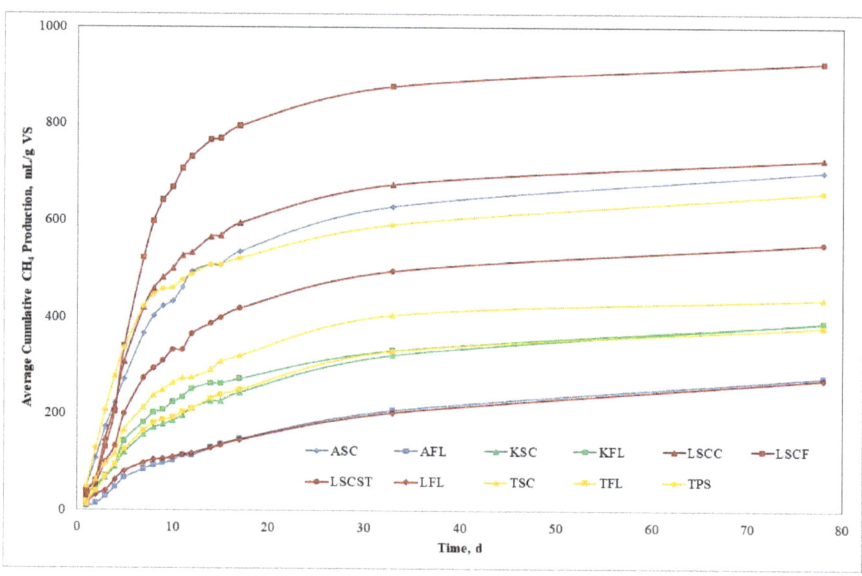

Figure 3. Biomethane production potential for the various samples collected from the four WWTPs of the study.

Table 2. Total net methane production in L/kg vs. for each sample.

Sample	Total Net CH$_4$ (L/kg VS)	Standard Deviation
ASC	522.2810	38.9924
AFL	95.0346	59.6485
KSC	207.4496	57.2870
KFL	207.4093	9.4035
LSCC	544.3497	27.5496
LSCF	741.9625	65.2636
LSCST	364.8005	44.7425
LFL	89.9468	14.4040
TSC	255.7938	54.7417
TFL	200.3265	32.0558
TPS	472.7396	48.0594
I	179.0224	3.8975

A crucial parameter in the efforts for the energy utilization of WWTPs residues is the time required to deliver an adequate amount of biogas, in relation to existing anaerobic fermenters of secondary sludge corresponding to sludge retention times of about 30 days. The kinetics of methane production for the various samples are shown in Figure 4, while the calculated maximum specific growth rate μ_{max} and methanogens doubling times are given in Figure 5.

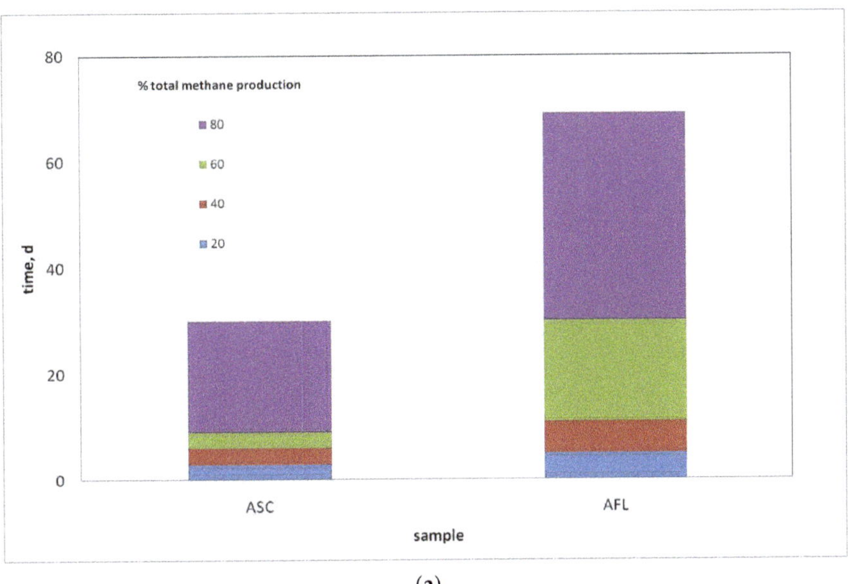

(**a**)

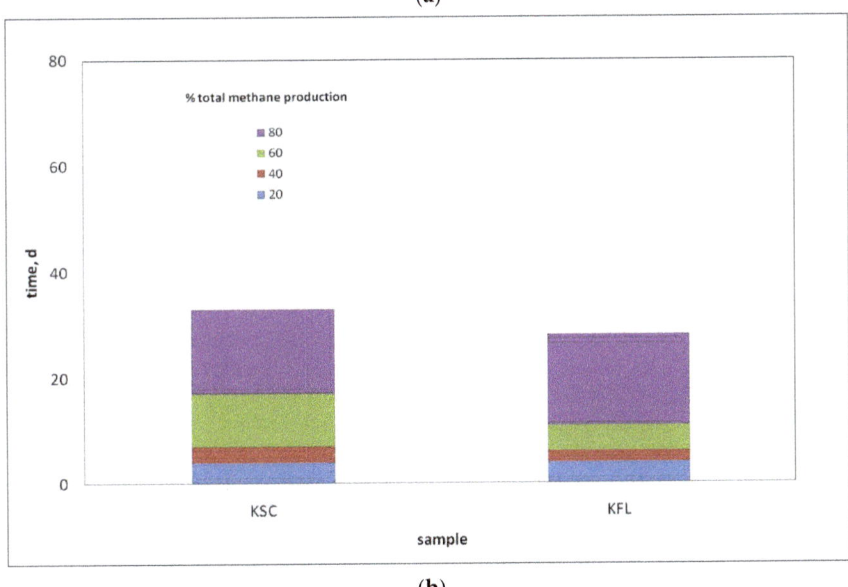

(**b**)

Figure 4. *Cont.*

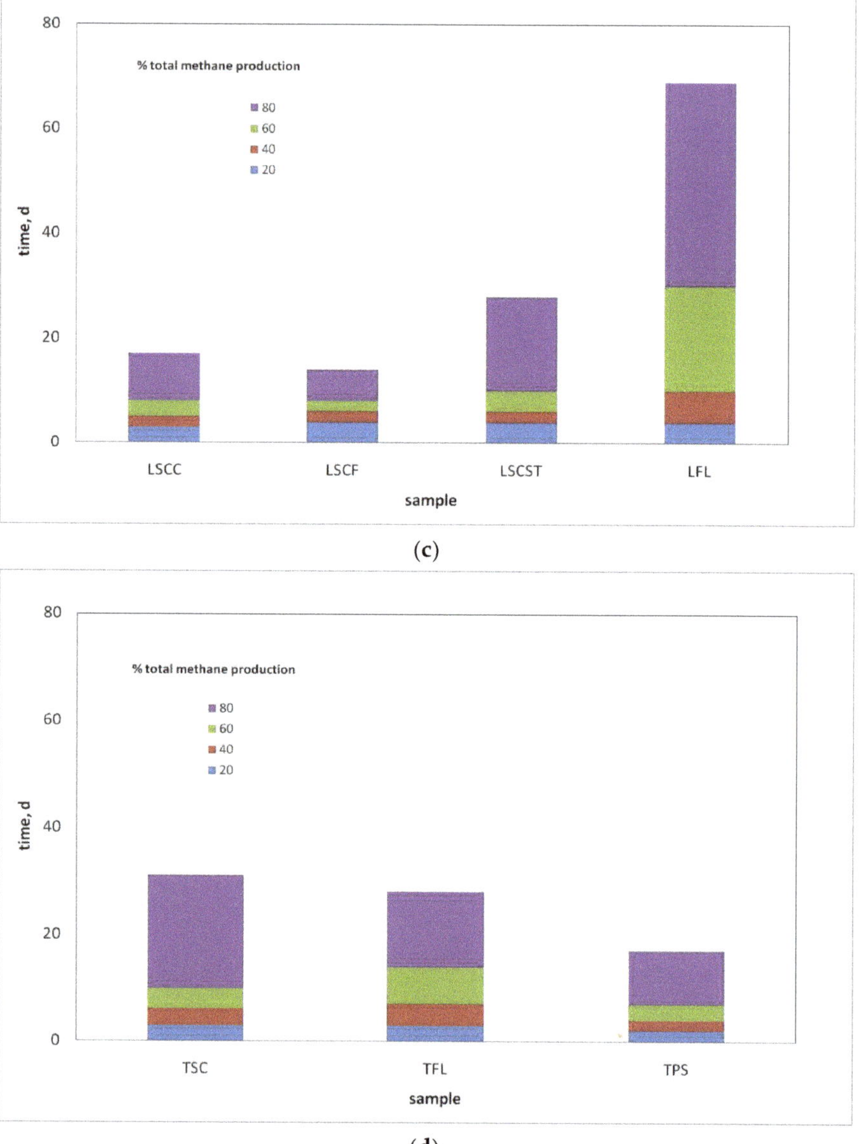

(**c**)

(**d**)

Figure 4. Time required for the production of 20, 40, 60, and 80% percentage vol. of total methane for the samples collected from WWTP No 1 (**a**), 2 (**b**), 3 (**c**), 4 (**d**).

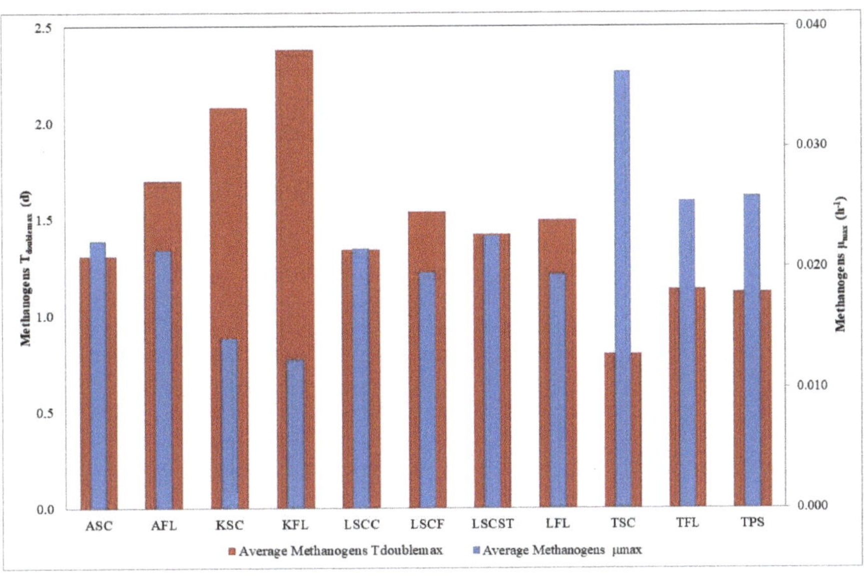

Figure 5. Estimated µ$_{max}$ values and average methanogens doubling time as deduced from BMP curves of various samples.

4. Discussion

The aim of the study is the determination of the valorization potential of byproducts produced during the primary treatment of municipal wastewaters by the utilization of their energy content through anaerobic digestion. The BMP potential of the individual byproducts was utilized as an indication of their capacity for production of biogas; this parameter is commonly used for the estimation of the theoretical maximum potential of various wastes [26]. In addition, BMP represents a required preliminary step for the examination of the feasibility of anaerobic digestion of a wide range of wastes and products, and it has been applied in this study to elucidate the assessment of the properties of these unconventional energy sources. Such an approach will in addition reveal potential operation problems during the addition of screenings in sludge digestion, such as inhibition of the anaerobic biota resulting in negligible biogas production. On the other hand, successful results will enhance byproducts utilization, and expected potential benefits include the reduction of side streams amounts conventionally disposed in sanitary landfills, as well as suppression of aeration demands in the following activated sludge process in a wastewater treatment plant. Nevertheless, the amount of these byproducts favors their co-digestion in existing sewage sludge anaerobic digestion units rather than their treatment in individual reactors. Their co-digestion is expected to result in raising the biogas production rate compared to sludge treatment solely, accounting for about 300–400 mL/g vs. [27]. Additional effects on digestate quality should be encountered, as soon as the feasibility of the process will be justified through the estimation of BMP potential. Nevertheless, a study has been reported on the quality of digestate from anaerobically treated screenings which was similar to the corresponding one from sewage sludge, towards its valorization through phosphorous recovery [17]. The overall estimation of potential economic and energy revenues due to the introduction of these side streams in existing treatment plants requires an integrated assessment of all potential benefits expected, including energy earnings from their utilization, reduction of landfilling costs, environmental costs revenues associated with fewer total GHG emissions, etc.

According to the information provided in Table 1, samples were collected from four WWTPs, with a wide range of design capacity, and servicing areas of different human activities. Selected WWTPs represented processes with a wide range of wastewater flowrates, ranging from as low as 8400 m^3/d to values exceeding 155,000 m^3/d. In addition, these units receive wastewaters from septic tanks reflecting, therefore, an additional contribution to screenings production. On the other hand, the composition of screenings was expected to vary due to the different origin of influents.

The lowest values of volatile solids were observed for the screenings collected during delivery of effluents from septic tanks transferred to one treatment plant by trucks and for the sewage screenings and the primary sludge screenings of the same WWTP. It seems that the volatile content of the samples can be attributed to their origin and the treatment stage: wastewater in septic tanks remains for a long time, subjected to anaerobic degradation of organic matter. Therefore, volatile solids in these effluents are expected to be lower than the corresponding solids removed in sewage treatment systems. On the other hand, the low volatile content of the screenings from the treatment plant with the highest capacity might be associated with the long sewage network system of the certain effluents, and their large transfer time from point of source to the treatment plant which might result in less organics content. Nevertheless, the fat content in the samples presented values reaching up to 30% of the total dry content, with great variation between the samples collected even from the same treatment plant. Similar values of volatile solids content have been reported in the literature ranging from 77 up to 95% of dry solids, depending on the screen size and the wastewater origin, with raw samples exhibiting considerably great variations in the moisture and the total solids content [11,17].

In addition to solids content, according to data given in Figure 2, high leached phosphorous and ammonia nitrogen contents were measured for the samples collected from the WWTP receiving effluents from an area with combined human activities, including extended agricultural areas and farming lands. Nevertheless, low leached organic matter values were observed in these samples, while the highest values of COD, up to 1000 mg/L, were measured for the other samples, without, however, an indication of potential trends due to the origin or wastewater characteristics. Limited studies related to the content of leached compounds from screenings have been reported, citing rather high COD values, ranging from 0.8 up to 1.6 g COD/g VS, close to theoretically estimated ones [28], while most studies include the elemental analysis of carbon, nitrogen, and phosphorous in screenings [13]. It must be underlined that biogas production potential is favored under certain ratios of organic matter: nitrogen:phosphorous and the presence of appropriate amounts of both carbon and nutrients sources is required for the efficient operation of anaerobic biocommunities. Although elemental nitrogen and phosphorous concentrations were not measured in this work, the corresponding leached contents of the nutrients were determined, as an indication of their presence in appropriate amounts and their availability towards the efficient operation of the anaerobes.

BMP results are given in Figure 3 as a function of reaction time, while the corresponding total net methane production is provided in Table 2. The methane potential is defined as the maximum produced methane of a specific substrate [29]; however, a methane production curve often presents an initial lag phase, and, therefore, the methane potential measurement for 80 days, as used in the study, provides valuable information about the requirement of a pretreatment step to speed up the whole process. As shown in Table 2, all samples exhibited methane production ranging from low values of around 90 L/kg vs. up to 740 L/kg vs. observed for fine screenings. Surprisingly, floatings from grit chambers had the lowest BMP of all samples from each wastewater treatment plant, although they were expected to deliver high biogas production potential due to their high content in fats (Figure 1) and the corresponding process stage where these samples are collected. As can be seen in Figure 1, fats represent a small fraction of the organic matter, and it is assumed that organic compounds other than fats in these samples were not easily assimilated by the anaerobes towards biogas production. In addition, it has been reported that the complicated composition of floatings containing low density materials such as cellulosic fibers can greatly affect the corresponding energy utilization available for anaerobic digestion [19], while methanogens can convert only a fraction of organics to biogas ranging from 50 to 60% [10,30]. On the other hand, methane production has been reported to be inhibited due to the formation of long-chain fatty acids during fats anaerobic degradation [31]. Nevertheless, the low BMP potential of floatings from the other two treatment plants is in line with their low content of fats.

Screenings from the plants located in urban areas were similar or slightly higher than the corresponding BMP potential of floatings, while large differences were observed in BMP values of screenings collected from treatment plants receiving effluents from areas with combined activities. The highest methane potential values were measured for the fine and the coarse screening samples from the plant with agro-industrial activities. This may be attributed to various reasons, such as the short distance of the sewage network and the corresponding short time of effluents travelling to the treatment plant, which is allowing less washing and degradation of the compounds, or the presence of potential effluents from agricultural and farming lands in the municipal effluents with a high methane production capacity. Moreover, samples originating from the primary sludge screen presented a rather high methane potential, the highest of the samples obtained from the certain treatment plant. This is a valuable indication of the high energy content of that sludge which is currently disposed together with the secondary anaerobically treated sludge in composting units, with loss of the contained energy content. Efforts were taken to identify potential relations of BMP values shown in Figure 3 to the corresponding chemical composition of the samples in Figures 1 and 2, but no significant trends were observed.

It should be mentioned that substrates obtained from wastes and residues of municipal wastewater treatment plants may have considerable methane potential, depending on their origin and the wastewater treatment stage: anaerobic digestion of FOGs resulted in the production of 271 to 344 and from 325 to 681 mL of methane per g of vs. added, corresponding to samples with volatile solid content up to 30% and 60%, respectively [20]. Similar results regarding methane production capacity were observed by screenings fed to a continuous anaerobic bioreactor ranging from 197–512 mL of CH_4/g vs. with 41.5% total and 29.4% volatile solids content [32]. Nevertheless, wastewaters from livestock production represent more efficient substrates, as they contain higher amounts of total solids and fatty compounds per m^3 of raw material than municipal wastewater: The theoretical biomethane potential from fats is estimated to be about 990 mL/g, while that from proteins is about 640 mL/g vs. and from hydrocarbons is about 415 mL/g vs. [33]. Although residues from wastewater treatment plants may have lower biogas potential than wastes of livestock production, the benefits from the utilization of the energy content is profound, considering the current methods of handling, including either disposal in sanitary landfill or incineration in municipal facilities.

Data in Figure 4 indicate the time required to receive a certain volume percentage of the total methane production, i.e., 20, 40, 60, and 80%, for each sample. The wider a column, the longer the time required to produce the corresponding methane volume, while shorter columns and less wide bars represent samples producing methane in short times. Similar observations can be drawn from the results shown in Figure 5, where μ_{max} and methanogens doubling time are reverse parameters, the latter corresponding to the time required for doubling the cells of the anaerobic micro-organisms. The lower the methanogens doubling time, the higher the maximum growth rate μ_{max}.

As shown in Figure 4, more than 50% of gas volume is produced at short reaction times, while production of up to 80% of total gas required longer time, depending on the particular sample. Samples collected from the WWTPs with low design capacities required rather long times to reach up to 80% volume of methane, exceeding 70 days, while the shortest times were observed for the residues from the WWTP in the agricultural area, varying between 15 and 30 days (although floatings may require as long as 70 days). Moreover, 60% of total gas volume was already produced in a period of less than 10 days for these samples, and a similar behavior was observed for the corresponding samples from the high-capacity treatment plant. However, slow kinetics were obtained for the residues from the other plants.

These results are in line with the corresponding data on average methanogens doubling time given in Figure 5, where short times were calculated for the high design capacity unit, ranging from 0.8 to 1.13 days, followed by the agricultural area treatment plant with slightly higher methanogens doubling times from 1.34 to 1.53 days. The highest methanogens doubling times were measured for the low-capacity plant with semi-industrial characteristics, reaching up to 2.38 days, almost three times higher than the lowest value. The above observations are becoming important for the energy recovery of these substrates during co-digestion with secondary sludge in these units. The low methanogens doubling time is associated with short retention times for the treatment of these substrates, which could strongly affect the design of anaerobic reactors in a new WWTP by requiring smaller reactor volumes. In addition, the range of the μ_{max} values and consequently of the doubling time is within that reported for a wide range of samples [22,34], and, therefore, the methanogens' growth rate was not differentiated due to the particular substrates used in this study.

Taking into account the BMP results presented in Table 2 and Figure 3 and the corresponding kinetic data in Figures 4 and 5, the beneficial role of the corresponding samples as potential energy sources could be identified. In general, all samples represent promising materials for energy recovery and the enhancement of energy production processes in the corresponding wastewater treatment plants. Nevertheless, efforts to identify the samples with the highest potential for biogas production should take into consideration both biogas cumulative volume and biota doubling time: under that framework, the fine and coarse screenings from the wastewater treatment plant receiving effluents from areas with mixed activities represent the best candidates for energy recovery through anaerobic digestion, followed by the primary sedimentation sludge screenings and the coarse screenings of the high flowrate municipal treatment plant. Both plants have already under operation an anaerobic digestion plant of secondary sludge, and the proposed energy utilization of their primary residues might represent an easy-to-apply process aiming at their energy self-efficiency.

Author Contributions: Conceptualization, P.S. and T.A.K.; methodology, A.M., I.L., C.A.T., and E.P.T.; validation, S.D.K., T.A.K., and P.S.; writing—original draft preparation, C.A.T. and S.D.K.; writing—review and editing, S.D.K., T.A.K., and P.S.; supervision, T.A.K. and P.S.; project administration, P.S. All authors have read and agreed to the published version of the manuscript.

Funding: This research was co-financed by the European Regional Development Fund of the European Union and Greek national funds through the Operational Program Competitiveness, Entrepreneurship and Innovation, under the call RESEARCH—CREATE—INNOVATE (project code T1EDK-00284).

Institutional Review Board Statement: Not applicable.

Informed Consent Statement: Not applicable.

Data Availability Statement: The data presented in this study are available on request from the corresponding author.

Acknowledgments: This research was co-financed by the European Regional Development Fund of the European Union and Greek national funds through the Operational Program Competitiveness, Entrepreneurship and Innovation, under the call RESEARCH—CREATE—INNOVATE (project code T1EDK-00284).

Conflicts of Interest: The authors declare no conflict of interest.

References

1. Gu, Y.; Li, Y.; Li, X.; Luo, P.; Wang, H.; Robinson, Z.P.; Wang, X.; Wu, J.; Li, F. The feasibility, challenges of energy self-sufficient wastewater treatment plants. *Appl. Energy* **2017**, *204*, 1463–1475. [CrossRef]
2. Gu, Y.; Li, Y.; Li, X.; Luo, P.; Wang, H.; Wang, X.; Wu, J.; Li, F. Energy self-sufficient wastewater treatment plants: Feasibilities, challenges. *Energy Procedia* **2017**, *105*, 3741–3751. [CrossRef]

3. Henriques, J.; Catarino, J. Sustainable value–An energy efficiency indicator in wastewater treatment plants. *J. Clean. Prod.* **2017**, *142*, 323–330. [CrossRef]

4. Panepinto, D.; Fiore, S.; Zappone, M.; Genon, G.; Meucci, L. Evaluation of the energy efficiency of a large wastewater treatment plant in Italy. *Appl. Energy* **2016**, *161*, 404–411. [CrossRef]

5. Statistical Office of the European Union; Eurostat. *Simplified Energy Balances-2015 Annual Data: 2017 Edition*; Tech Report; Eurostat: Luxemburg, 2017.

6. Papa, M.; Foladori, P.; Guglielmi, L.; Bertanza, G. How far are we from closing the loop of sewage resource recovery? A real picture of municipal wastewater treatment plants in Italy. *J. Environ. Manag.* **2017**, *198*, 9–15. [CrossRef]

7. Shizas, I.; Bagley, D.M. Experimental determination of energy content of unknown organics in municipal wastewater streams. *J. Energy Eng.* **2004**, *130*, 45–53. [CrossRef]

8. Heidrich, E.S.; Curtis, T.P.; Dolfing, J. Determination of the internal chemical energy of wastewater. *Environ. Sci. Technol.* **2011**, *45*, 827–832. [CrossRef]

9. Kales, M.; Palmowski, L.; Pinnekamp, J. Carbon recovery from screenings for an energy efficient wastewater treatment. *Water Sci. Technol.* **2017**, *76*, 3299–3306. [CrossRef] [PubMed]

10. Cadavid-Rodriguez, L.S.; Horan, N. Methane production, hydrolysis kinetics in the anaerobic degradation of wastewater screenings. *Water Sci. Technol.* **2013**, *68*, 413–418. [CrossRef]

11. Hyaric, R.L.; Canler, J.-P.; Barillon, B.; Naquin, P.; Gourdon, R. Characterization of screenings from three municipal wastewater treatment plants in the Region Rhone-Alpes. *Water Sci. Technol.* **2009**, *60*, 525–531. [CrossRef]

12. Statistisches Bundesamt DESTATIS Statistisches Bundesamt (Hrsg.). *Umwelt, Abfallentsorgung 2008*; Fachserie 19 Reihe 1; DESTATIS: Wiesbaden, Germany, 2010.

13. Paulsrud, B.; Rusten, B.; Aas, B. Increasing the sludge energy potential of wastewater treatment plants by introducing fine mesh sieves for primary treatment. *Water Sci. Technol.* **2014**, *69*, 560–565. [CrossRef]

14. Razafimanantsoa, V.A.; Ydstebø, L.; Bilstad, T.; Sahu, A.K.; Rusten, B. Effect of selective organic fractions on denitrification rates using Salsnes Filter as primary treatment. *Water Sci. Technol.* **2014**, *69*, 1942–1948. [CrossRef]

15. Kehrein, P.; van Loosdrecht, M.; Osseweijer, P.; Garfí, M.; Dewulf, J.; Posada, J. A critical review of resource recovery from municipal wastewater treatment plants—Market supply potentials, technologies, bottlenecks. *Environ. Sci. Water Res. Technol.* **2020**, *6*, 877–910. [CrossRef]

16. Abomohra, A.E.F.; Elsayed, M.; Esakkimuthu, S.; El-Sheekh, M.; Hanelt, D. Potential of fat, oil, grease (FOG) for biodiesel production: A critical review on the recent progress, future perspectives. *Prog. Energy Combust. Sci.* **2020**, *81*, 100868. [CrossRef]

17. Wid, N.; Horan, N.J. Anaerobic digestion of wastewater screenings for resource recovery, waste reduction. *IOP Conf. Ser. Earth Environ. Sci.* **2016**, *36*, 12017. [CrossRef]

18. Hyaric, R.L.; Canler, J.-P.; Barillon, B.; Naquin, P.; Gourdon, R. Pilot-scale anaerobic digestion of screenings from wastewater treatment plants. *Bioresour. Technol.* **2010**, *101*, 9006–9011. [CrossRef]

19. Ruiken, C.J.; Breuer, G.; Klaversma, E.; Santiago, T.; van Loosdrecht, M.C.M. Sieving wastewater e Cellulose recovery, economic, energy evaluation. *Water Res.* **2013**, *47*, 43–48. [CrossRef]

20. Long, J.H.; Aziz, T.N.; Reyes, F.L.; Ducoste, J.J. Anaerobic co-digestion of fat, oil, grease (FOG): A review of gas production, process limitations. *Process. Saf. Env. Prot.* **2012**, *90*, 231–245. [CrossRef]

21. Rice, E.W.; Baird, R.B.; Eaton, A.D. *Standard Methods for the Examination of Water, Wastewater*, 23rd ed.; American Public Health Association; American Water Works Association; Water Environment Federation: Washington, DC, USA, 2017; Volume 1.

22. Kalamaras, S.D.; Vasileiadis, S.; Karas, P.; Angelidaki, I.; Kotsopoulos, T.A. Microbial adaptation to high ammonia concentrations during anaerobic digestion of manure-based feedstock: Biomethanation, 16S rRNA gene sequencing. *J. Chem. Technol. Biotechnol.* **2020**, *95*, 1970–1979. [CrossRef]

23. Angelidaki, I.; Alves, M.; Bolzonella, D.; Borzacconi, L.; Campos, J.L.; Guwy, A.J.; Kalyuzhnyi, S.; Jenicek, P.; Van Lier, J.B. Defining the biomethane potential (BMP) of solid organic wastes, energy crops: A proposed protocol for batch assays. *Water Sci. Technol.* **2009**, *59*, 927–934. [CrossRef] [PubMed]

24. Fotidis, I.A.; Karakashev, D.; Angelidaki, I. Bioaugmentation with an acetate-oxidising consortium as a tool to tackle ammonia inhibition of anaerobic digestion. *Bioresour. Technol.* **2013**, *146*, 57–62. [CrossRef]

25. Khoshnevisan, B.; Tsapekos, P.; Zhang, Y.; Valverde-Pérez, B.; Angelidaki, I. Urban biowaste valorization by coupling anaerobic digestion, single cell protein production. *Bioresour. Technol.* **2019**, *290*, 121743. [CrossRef]

26. Da Silva, C.; Astals, S.; Peces, M.; Campos, J.L.; Guerrero, L. Biochemical methane potential (BMP) tests: Reducing test time by early parameter estimation. *Waste Manag.* **2018**, *71*, 19–24. [CrossRef]

27. Davidsson, Å.; Lövstedt, C.; la Cour Jansen, J.; Gruvberger, C.; Aspegren, H. Co-digestion of grease trap sludge, sewage sludge. *Waste Manag.* **2008**, *28*, 986–992. [CrossRef] [PubMed]

28. Cadavid-Rodriguez, L.S.; Horan, N. Reducing the environmental footprint of wastewater screenings through anaerobic digestion with resource recovery. *Water Environ. J.* **2012**, *26*, 301–307. [CrossRef]

29. Hansen, T.L.; Schmidt, J.E.; Angelidaki, I.; Marca, E.; la Cour Jansen, J.; Mosbæk, H.; Christensen, T.H. Method for determination of methane potentials of solid organic waste. *Waste Manag.* **2004**, *24*, 393–400. [CrossRef]

30. Francioso, O.; Rodriguez-Estrada, M.T.; Montecchio, D.; Salomoni, C.; Caputo, A.; Palenzona, D. Chemical characterization of municipal wastewater sludges produced by two-phase anaerobic digestion for biogas production. *J. Hazard. Mater.* **2010**, *175*, 740–746. [CrossRef] [PubMed]
31. Cirne, D.G.; Paloumet, X.; Bjornsson, L.; Alves, M.M.; Mattiasson, B. Anaerobic digestion of lipid-rich waste—Effects of lipid concentration. *Renew. Energy* **2007**, *32*, 965–975. [CrossRef]
32. Kabouris, J.C.; Tezel, U.; Pavlostathis, S.G.; Engelmann, M.; Dulaney, J.; Gillette, R.A.; Todd, A.C. Methane recovery from the anaerobic codigestion of municipal sludge, FOG. *Bioresour. Technol.* **2009**, *100*, 3701–3705. [CrossRef]
33. Alves, M.M.; Pereira, M.A.; Sousa, D.Z.; Cavaleiro, A.J.; Picavet, M.; Smidt, H.; Stams, A.J. Waste lipids to energy: How to optimize methane production from long-chain fatty acids (LCFA). *Microb. Biotechnol.* **2009**, *2*, 538–550. [CrossRef]
34. Fotidis, I.A.; Karakashev, D.; Kotsopoulos, T.A.; Martzopoulos, G.G.; Angelidaki, I. Effect of ammonium, acetate on methanogenic pathway, methanogenic community composition. *FEMS Microbiol. Ecol.* **2013**, *83*, 38–48. [CrossRef] [PubMed]

 *sustainability*

Article

The Effect of Thermal Treatment on the Physicochemical Properties of Minerals Applied to Heterogeneous Catalytic Ozonation

Savvina Psaltou [1], Efthimia Kaprara [2], Kyriaki Kalaitzidou [2], Manassis Mitrakas [2] and Anastasios Zouboulis [1,*]

[1] Laboratory of Chemical and Environmental Technology, Department of Chemistry, Aristotle University, 54124 Thessaloniki, Greece; spsaltou@chem.auth.gr

[2] Laboratory of Analytical Chemistry, Department of Chemical Engineering, Aristotle University, 54214 Thessaloniki, Greece; kaprara@auth.gr (E.K.); kikalaitz@geo.auth.gr (K.K.); mmitraka@auth.gr (M.M.)

* Correspondence: zoubouli@chem.auth.gr

Received: 30 October 2020; Accepted: 14 December 2020; Published: 15 December 2020

Abstract: In order to enhance the efficiency of heterogeneous catalytic ozonation, the effect of thermal treatment on three commonly used and inexpensive minerals, i.e., zeolite, talc and kaolin (clay), which present different physicochemical properties as potential catalysts, has been examined for the removal of para-chlorobenzoic acid (p-CBA). p-CBA is considered a typical micro-pollutant, usually serving as an indicator (model compound) to evaluate the production of hydroxyl radicals in ozonation systems. The catalytic activity of selected solid catalysts was studied for different pH values (6, 7 and 8) and different temperatures (15 °C, 25 °C and 35 °C). The mechanism of radicals' production was also verified by the addition of tert-butyl alcohol (TBA). The respective thermal behavior study showed that the point of zero charge (PZC) of these minerals increased with the increase of applied treatment temperature, as it removed crystalline water and hydroxyls, thus improving their hydrophobicity. Circa-neutral surface charge and the presence of hydrophobicity were found to favor the affinity of ozone with solid/catalytic surfaces and the subsequent production of hydroxyl radicals. Therefore, zeolite and talc, presenting PZC 7.2 and 6.5 respectively, showed higher catalytic activity after thermal treatment, while kaolin with PZC equal to 3.1 showed zero to moderate catalytic efficiency. The degradation level of p-CBA by oxidation was favored at 25 °C, while the pH value exerted positive effects when it was increased up to 8.

Keywords: heterogeneous catalytic ozonation; PZC; p-CBA; minerals; thermal treatment; micropollutants removal

---

1. Introduction

Nowadays, persistent organic compounds (POPs) are considered chemical compounds of great environmental concern [1]. Several classes of them occur in aquatic ecosystems [2] but with rather low concentrations, ranging from ng/L–µg/L (they are thus termed as micropollutants). They are very stable compounds and cannot be easily degraded or removed by the application of conventional biological wastewater treatment technologies [1]. Although their quantities are still quite low, they can constantly accumulate in fragile ecosystems, and there is a high probability of side effects in living organisms throughout the food chain [2]. Municipal wastewater treatment plants have been identified as one of the main sources of pollution from micropollutants. The treated wastewaters may contain several compounds of emerging concern, released into water bodies and eventually ending up in marine environments [3].

Among the available degradation techniques that have been examined for the removal/oxidation/destruction of micropollutants is the process of ozonation. This treatment method is based on the oxidation of organic compounds through the presence of ozone molecules, as well as through the secondary formation of hydroxyl radicals. Hydroxyl radicals are produced in this case by ozone self-decomposition and present higher oxidative potential than the application of single ozonation. Additionally, their ability to react unselectively with different organic molecules makes them an (almost) ideal oxidizing agent for Advanced Oxidation Processes (AOPs). However, in order to further increase the efficiency of this oxidation system, it is important to accelerate the production of hydroxyl radicals. The addition of an appropriate catalyst can enhance ozone decomposition, leading to a catalytic ozonation process, which is a promising treatment technique for the removal of refractory organic compounds, such as micropollutants, from water and wastewater sources. In heterogeneous catalytic ozonation, appropriate solid materials are commonly used as catalysts, improving hydroxyl radical's production [4].

In this procedure, the effective contact between ozone molecules and the catalyst surface is very important, and to some extent it can be estimated by the respective Point-of-Zero-Charge (PZC) of the solid material [5]. PZC is generally described as the pH, where the net charge of the particle surface is equal to zero [6]. The overall charge of hydroxyl groups on the surface of a solid is responsible for the adsorption and subsequent decomposition of ozone. Hydroxyl groups can be negatively, positively or even neutrally charged according to the relevant solution's pH [5]. Based on several publications, it has been concluded that the desired decomposition of ozone into hydroxyl radicals is favored when the material/catalyst surface is neutrally charged [5,7–12]. Due to its structure, the ozone molecule presents both nucleophilic and electrophilic properties. The (electrophilic) hydrogen atoms and the (nucleophilic) oxygen atoms of hydroxyl groups can simultaneously react with ozone molecules, creating a ring when the hydroxyls are in a natural state. However, this ring is unstable and can easily break down into superoxide radicals ($O_2^{\bullet-}$), hence promoting the production of hydroxyl radicals. When hydroxyl groups are positively or negatively charged, the interaction of the surface with ozone molecules decreases and mass transfer between them is inhibited [5].

Wang et al. [7] and Zhu et al. [8] prepared two different magnesium nano-oxides (nano-MgO) with point-of-zero-charge values 10.2 and 7.2, respectively, and used them as catalysts for the removal of phenol and quinolone via the application of a heterogeneous catalytic ozonation process. However, both solid materials showed similar behavior due to their degradation/dissolution in water during the process. The contribution of adsorption to the overall process can be considered as rather negligible. Higher catalytic activity was observed when the catalyst's surface was neutrally charged (i.e., at pH=PZC). In other relevant studies, pumice has been used as a catalyst, either as a raw material or after its modification. Its point-of-zero-charge was 6-7, either as pumice [9] or after iron silicate [10] or iron [11] deposition in its structure. These studies also concluded that at pH 6-7, pumice presented the best catalytic action because it was neutrally charged. Zhao et al. [12] also tested Cu-cordierite material for the removal of nitrobenzene by catalytic ozonation. Nitrobenzene is a compound that practically cannot be removed by oxidation with molecular ozone, similar to the p-CBA used in this study. Due to the fact that the PZC was 6.94, it exhibited the highest catalytic activity at circa-neutral pH value, where the surface of the catalyst was also neutrally charged.

Sui et al. [13] used ferric oxy-hydroxide(FeOOH) as a catalyst for the removal of oxalic acid, with the adsorption of this pollutant in the catalyst surface being rather negligible. This research differed because the ozone decomposition and the subsequent production of hydroxyl radicals were found to be enhanced in both the neutral and the positive state of the material, i.e., at neutral and acidic pHs, respectively (PZC = 7.2). Similar results were published by Kermani et al. [14] and Liu et al. [15]; in both cases, positively charged materials showed the highest catalytic activity. Huang et al. [16] claimed that the oxygen groups on the catalyst surface were coupled with bisphenol A via hydrogen bonds, and thus the degradation of ozone was promoted, resulting in increased system efficiency. Through this mechanism, the removal of this micropollutant was increased by increasing the pH value.

Furthermore, the hydrophobic molecules of ozone can be attracted by non-polar surfaces, and thus the contact between them is favored [17]. By enhancing the hydrophobic degree of a material, the respective contact can be improved, and the production of hydroxyl radicals can be accelerated. A possible way in which both hydrophobic behavior and point-of-zero-charge (PZC) can be simultaneously increased is the thermal treatment of a material. During thermal treatment, hydrophilic groups, such as hydroxyls, are gradually removed. However, at elevated temperatures (i.e., >900 °C), the surface becomes more hydrophilic due to the decrease in other hydrophobic components, such as in the case of talc siloxanes [18].

The objectives of this work were to study the effects of thermal treatment on the major physicochemical properties of selected natural minerals, acting as potential catalysts in the ozonation process. Subsequently, these properties were correlated with the respective catalytic activity regarding the oxidation of p-CBA by the application of heterogeneous catalytic ozonation. The effect of applied common experimental conditions on their catalytic activity was also investigated. p-CBA was selected as a typical model organic compound because it cannot practically react with ozone molecules($k_{O3} < 0.15\ M^{-1}s^{-1}$), and it can only be degraded by the presence of hydroxyl radicals($k_{\bullet OH} = 5 \times 10^9\ M^{-1}s^{-1}$) [19]. Therefore, acceleration of their production can be indirectly estimated by the enhancement of p-CBA removal.

2. Materials and Methods

2.1. Materials

All the chemicals used in the experiments were of analytical grade. The acetonitrile and phosphoric acid used for p-CBA determination were HPLC-grade and purchased from Sigma-Aldrich (St. Louis, MO, USA). p-CBA was obtained from the same company, and it was used as a model compound at an initial concentration of 4 µM. All aqueous solutions were prepared in deionized water. For pH adjustment, K_2HPO_4 and KH_2PO_4 (Chem-Lab, Zedelgem, Belgium) were used for the preparation of a convenient buffer solution. All the examined solid materials/catalysts were commercially available, and they were pre-treated with ozone before the experiments. Zeolite (Clinoptilolite) from the Metaxades area (Alexandroupoli, Thrace region, Greece)and talc from the Arnissa area (Edessa, Macedonia region, Greece)were calcined at the temperatures 100 °C (zeolite-1, talc-1), 300 °C (zeolite-3, talc-3), 600 °C (zeolite-6, talc-6) and 800 °C (zeolite-8, talc-8), while kaolin (clay)from the Leukogia area (Drama, Macedonia region, Greece), which was used as a comparison measure, was only calcined at 600 °C and abbreviated as kaolin-6. Calcination was carried out by raising the temperature from room temperature up to 800 °C in 2 h, and then the materials were left to cool at room temperature.

2.2. Experimental Procedures

The ozonation experiments were performed in batch mode. O_3 was generated using a corona discharged O_3 generator (Ozonia-Triogen Model TOGC2A) with pure oxygen as the feed gas. The initial p-CBA concentration was 4 µM, and the pH values were adjusted between 6 and 8 with the addition of an appropriate amount of 0.005 M K_2HPO_4/KH_2PO_4 buffer solution. The initial concentration of O_3 was 2 mg/L, based on preliminary experiments. Solid materials/potential catalysts (0.5 g/L) were introduced to the dark reaction vessel just before the addition of O_3 solution. The catalytic ozonation reaction time was 30 min for all the experiments, and appropriate samples were received during convenient reaction time intervals, i.e., 1, 3, 15 and 30 min; filtered through 0.45 µm membrane filters; and the residual concentrations of O_3 and p-CBA were determined. The oxidation reaction was stopped by the addition of an appropriate amount of indigo solution. Single ozonation experiments were conducted in the same way but without the addition of a solid/catalyst, while adsorption experiments were performed without the addition of O_3. TBA was used as a scavenger to capture the $^\bullet OH$ radicals. The dosage of tert-butyl alcohol (TBA) purchased from Sigma-Aldrich was 0.3 mM, calculated based on the equation proposed by Wang et al. [20].

2.3. Analytical Techniques

Ozone concentrations were determined by the Indigo method [21]. The color change of indigo solution was measured at 600 nm by a spectrometer (Lange, DR3900). The specific surface area of catalysts was calculated by nitrogen gas adsorption at liquid N_2 temperature (77 K) by using anASAP2020 analyzer, according to the Brunauer–Emmet–Teller (BET) method. Point-of-zero-charge (PZC) values were determined by the immersion technique [22]. Aqueous solutions of $NaNO_3$ (0.1 M) with pH values ranging from 3 to 11 were prepared, and 0.25 g of the mineral was added to each solution. Suspensions were equilibrated for 24 h. The PZC value was identified as the pH value, where the pH of blank solution is equal to the pH of mineral suspension. A High-Performance Liquid Chromatography system (Thermo Fisher Scientific, Waltham, MA, USA, HPLC model of UV Spectrum UV2000) was used to determine the residual p-CBA concentration. An Agilent 4.6 × 250 mm reversed phase column (model Eclipse Plus C18) was used, and the compound was measured at 254 nm. The mobile phase consisted of acetonitrile and phosphoric acid (10 mM) in the ratio40:60 v/v. The pH of solutions was measured by a pH meter (Jenway, Cole Palmer, UK, model 3540). The mineralogical transformations of thermally treated materials were determined by X-ray diffraction (XRD), using a D8 Advanced instrument (Brucker, Karlsruhe, Germany) operating at 40 mA and 40 kV with Cu radiation (λ = 1.54 nm),the detector LYNXEYE (1D mode) (Brucker Karlsruhe, Germany) were used. The data were collected in the 2θ range of 5–60°, with increments of 0.010° and a counting time of 0.1 s per step for talc and 0.2 s for zeolite materials. The rotation time was 15 per min for both materials. Thermogravimetric analysis and Differential Thermal Analysis (TG-DTA) were obtained simultaneously by using a Perkin Elmer (Waltham, MA, USA) STA 6000 Thermal Analyzer instrument. The samples were heated from 20 to 900 °C at a constant rate of 20 °C/min in a nitrogen atmosphere (flow 20 cm^3/min). Regarding the chemical characterization of minerals, the total content of chemical components was determined by the application of flame atomic adsorption spectroscopy (Perkin Elmer, Waltham, MA, USA, model AAnalyst 800).

3. Results and Discussion

3.1. Catalysts Characterization

The first part of this study focuses on the effect of thermal treatment on the structure and major physicochemical characteristics of the examined minerals/potential catalysts for the heterogeneous catalytic oxidation process. Zeolite and talc were thermally treated at various temperatures up to 800 °C, while kaolin was only calcined at 600 °C. The respective specific surface area (S_{BET}) of the materials was rather low, e.g., 21 m^2/g, 10.5 m^2/g and 13 m^2/g for zeolite, talc and kaolin, respectively, and their chemical compositions are shown in Table 1.

Table 1. Chemical composition of examined raw minerals.

Minerals	Chemical Composition (%)								mg/Kg	
	SiO_2	Al_2O_3	MgO	CaO	Na_2O	K_2O	Fe_2O_3	LOI	NiO	Cr_2O_3
Zeolite	70.5	12.1	0.58	2.91	1.45	2.82	0.67	8.9 [a]	ND	ND
Talc	65.8	ND	28.6	0.01	0.01	0.04	0.73	4.6 [b]	1500	20
Kaolin	50.9	33.1	0.71	0.10	0.28	1.19	0.70	12.9 [a]	120	ND
Detection Limit	0.1	0.05	0.0005	0.004	0.001	0.002	0.01	0.1	100	50

[a]: Determination at 900 °C as clarified by TG-DTA. [b]: Determination at 1200 °C, since the loss on ignition of talk is partial at 900 °C as verified by TG-DTA.

Generally, the mineral structure contains adsorbed and structural water molecules. Under the effect of thermal treatment, the water molecules are removed at different stages. Thermogravimetric Analysis (TGA) was used to monitor the respective weight loss of minerals with the increase in temperature. The TGA diagram of zeolite (Figure 1a) shows continuous mass loss during heating up to 700 °C, due to the initial loss of hydroscopic water and the subsequent de-hydroxylation of the crystal lattice. These TGA results suggest that the surface chemistry of zeolite changes with thermal treatment. On the other hand, the diagram of talc (Figure 1b) shows that there was only a small weight loss due to the removal of adsorbed water molecules, even at the higher temperatures. According to Yi et al. [18], the de-hydroxylation of talc only occurs at temperatures higher than 900 °C, which were not part of the present study. The weight loss for talc up to 900 °C was only 1.3%, while in the same temperature range; zeolite lost 8.9% of its weight. This is evidence that in talc, no major structural modifications are occurring. During thermal treatment, both minerals only displayed an endothermic effect.

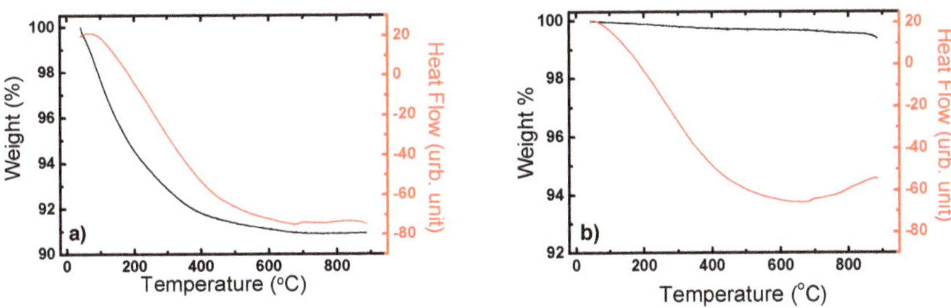

Figure 1. Analysis and Differential Thermal Analysis (TG-DTA) diagrams of (**a**) zeolite and (**b**) talc.

The XRD diagrams of raw materials, as well as of the thermally treated samples, are shown in Figure 2. Heating at 900 °C did not significantly affect talc's mineral structure. On the contrary, the zeolite material collapsed when heated at 600 °C, while at even higher temperatures it was transformed to poorly crystallized phases. Upon heating, zeolites undergo a series of chemical and structural changes until they are eventually largely converted to an amorphous material [23]. Similar observations have been reported by Christidis et al. [24], where a Greek zeolite collapses after heating at 450 °C. Clinoptilolite dehydrates at temperatures up to 300 °C and dehydroxylates at temperatures up to 800 °C, while at even higher temperatures, it decomposes and transforms into amorphous Al-Si and Si glasses [25].Figure 2a shows the initial raw zeolite containing clinoptilolite (card PDF 01-089-7539) as its major crystalline phase, accompanied by orthoclase (card PDF 01-086-0437), anorthoclase (card PDF 01-075-1634) and quartz (card PDF 01-079-1910). The same structure was presented by the respective calcined material at 300 °C. When zeolite was thermally treated above 500 °C, the characteristic reflections/peaks of clinoptilolite disappeared entirely, and only those of orthoclase, anorthoclase and quartz remained. Figure 2b shows the XRD spectra of talc before and after thermal treatment at temperatures of 300 °C, 600 °C and 800 °C. The diffraction peaks, appearing at 9.44°, 18.97°, 28,60° and 48.79°, agree well with the crystal structure of talc, indicating that both the raw material and the thermally treated material used in this work were talc.

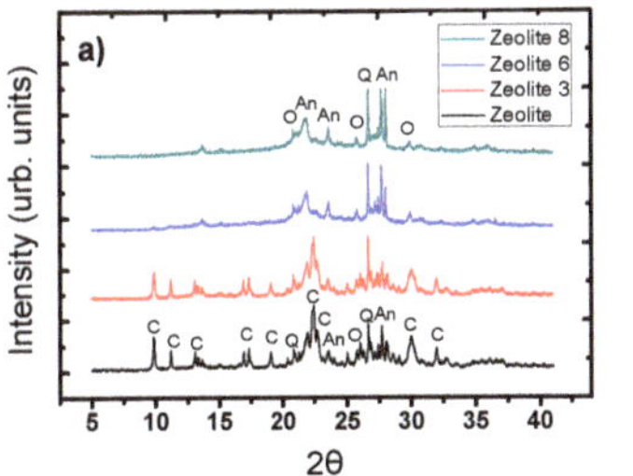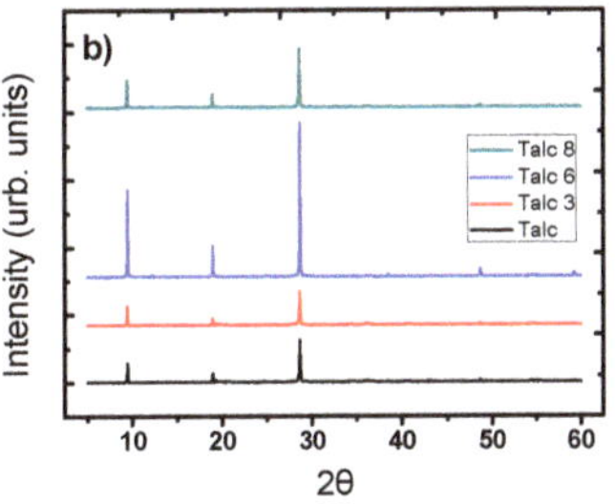

Figure 2. X-ray diffraction (XRD) patterns of raw materials and of thermally treated samples: (**a**) zeolite and (**b**) talc. Peaks for (**a**): C-clinoptilolite, O-orthoclase, An-anorthoclase, Q-quartz.

The surface charge of any solid material in a solution is based mainly on its PZC value. The PZC of raw kaolin was 3.1, and this value was unchanged up to the temperature of 100 °C. After that, there was an increase in the PZC value, probably due to dehydroxylation. At 800 °C, the PZC increased to 4.4, probably due to the formation of meta-kaolin, as Torres Sánchez and Tavani observed [6]. Its structural water began to be removed at 500 °C and completed at 600 °C, as verified by the weight loss in Supplementary Materials Figure S1. In the same temperature range, the PZC of kaolin increased sharply. The PZC of zeolite and of talc were also found to increase but to a lower extent. The PZC of raw zeolite and of talc were 6.8 and 5.9, and after calcination at 800 °C, they increased by 0.4 and 0.8 units, respectively. Changes to the values of PZC depend on the heating temperature of each material and are shown in Figure 3.

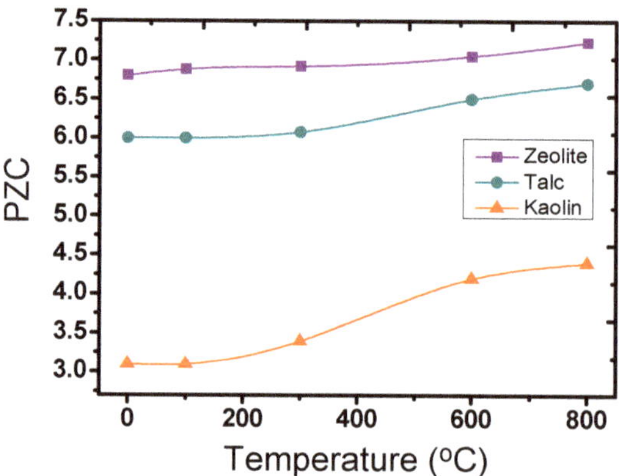

Figure 3. Point-of-zero-charge (PZC) of zeolite, talc and kaolin after calcination treatment in the temperature range 100–800 °C.

One of the main mechanisms regarding the formation of charges on the surface of an oxide in aqueous media is the sorption of metal hydrolytic complexes derived from the hydrolysis of solid material [26]. Therefore, the differences in PZC values in the experiments were provoked by the formation of these hydro-complexes on the surface of minerals. This is in accordance with the results of other relevant studies on several oxides [6,27]. Furthermore, Stanković et al. [28] have observed that the PZC of a hydrous oxide was found to depend more on the hydration degree than on the crystal structure of the oxide, and generally the PZC value shifts towards the alkaline pH region during the dehydration process.

3.2. Catalytic Activity of the Examined Natural Minerals

In this study, the catalytic activity of three natural minerals was investigated before and after their thermal treatment, regarding the application of heterogeneous catalytic ozonation for the destruction by oxidation of p-CBA. All these materials showed rather low uptake (adsorption) capacities independently of calcination temperature (Table S1). Therefore, none of them can be characterized as an efficient adsorbent material, at least for the examined micropollutant. The absorption capacity is illustrated in Figure 1 for the negative values of time.

The catalytic activity of these solids was evaluated through p-CBA removal efficiency in the ozonation process, and the obtained results of heterogeneous catalytic ozonation were compared with those of single ozonation application. The red, dashed line in Figure 4 and Figures 6–8 represents the p-CBA analytical detection limit. All these materials were found to enhance the decomposition of ozone when compared with single ozonation (Figure S2), but not all of them were found to present catalytic activity. Catalytic activity means the acceleration of p-CBA removal due to the enhancement of hydroxyl radical production. Raw zeolite and talc can be considered as catalysts, while kaolin, neither before or after its calcination, was found to present any catalytic activity (Figure 4). Kaolin is a material with PZC 3.1. After thermal treatment, the PZC of kaolin increased to 4.7. Therefore, in both forms, it is a strongly negatively charged solid in the neutral pH range, where most of the experiments were performed. Ozone molecules tend to approach neutrally charged and non-polar surfaces more effectively, enhancing their decomposition into hydroxyl radicals when coming into contact [5,7–9,29]. Raw talc (with PZC 5.9) presents lower catalytic activity than talc-8, which has PZC equal to 6.5and a more hydrophobic surface. The catalytic activity of neutrally charged raw zeolite was enhanced through the application of thermal treatment, because the PZC and the hydrophobicity of the material increased. The best relevant performance was presented by the zeolite-8, where the efficiency of treatment process was 98.5% even from the first min of reaction/oxidation time. All zeolites showed higher catalytic activity even than talc-8 because they were neutrally charged, while talc-8 with PZC 6.5 was slightly negatively charged throughout the oxidation process. However, among the talc materials, talc-8 was optimum and removed 95.5% of p-CBA, even from the first min of oxidation.

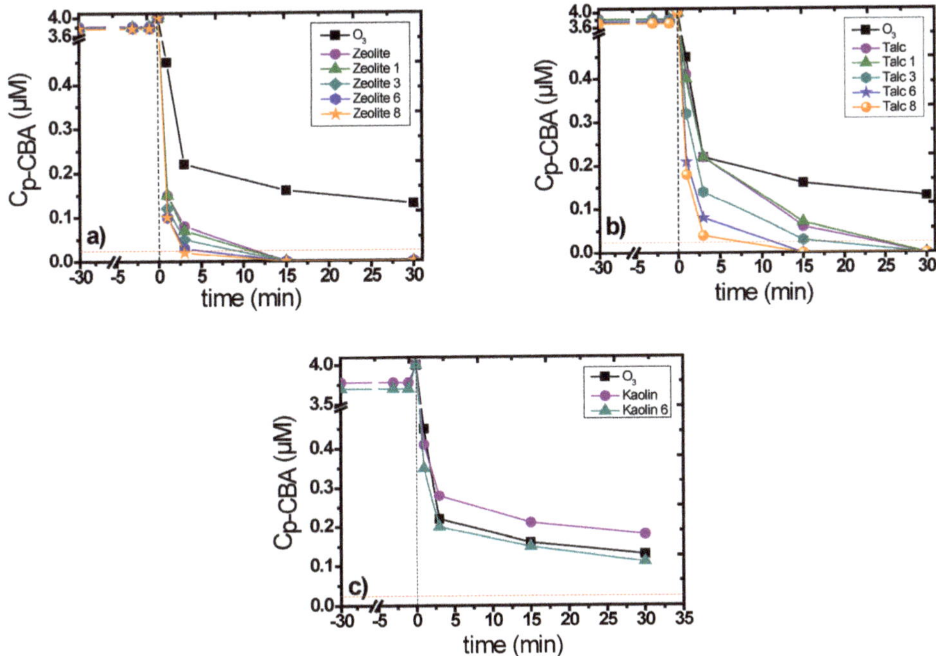

Figure 4. Effect of calcination temperature on the catalytic activity of examined natural minerals: (**a**) zeolite, (**b**) talc, and (**c**) kaolin; experimental conditions: $C_{p\text{-}CBA}$ 4 µM, C_{O3} 2 mg/L, $C_{cat.}$ 0.5 g/L, pH 7, temperature 23 ± 2 °C.

The high correlation between PZC values and the efficiency of the heterogeneous catalytic ozonation process is shown in Figure 5 for the cases of thermally treated zeolite and talc. An almost linear response was observed between PZC values and their respective catalytic ozonation efficiency. Correlation was performed considering the values from the first min of reaction/oxidation time because differences in respective performances at that early oxidation time were more pronounced. The difference in the efficiency between talc and talc-8 was higher when compared with relevant observations in the case of zeolite, because the increase of the PZC value was greater for talc (Figure 3).

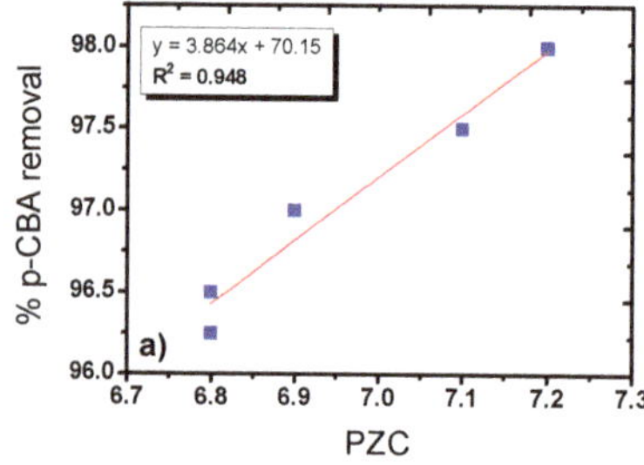
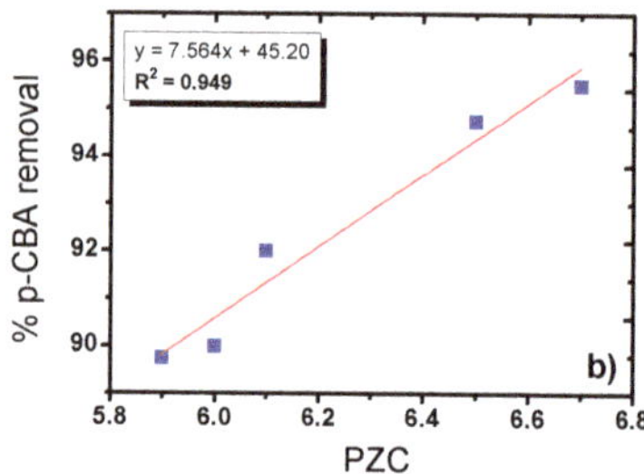

Figure 5. Correlation between PZC values (as obtained from raw and thermally treated mineral samples) and the respective catalytic efficiency calcinated materials, considering the first min of reaction/oxidation:(**a**) zeolite and (**b**) talc; experimental conditions: $C_{p\text{-}CBA}$ 4 µM, C_{O3} 2 mg/L, $C_{cat.}$ 0.5 g/L, pH 7, temperature 23 ± 2 °C.

In addition to removing/oxidizing p-CBA, other evidence that the catalytic ozonation process was based on the formation of the radicals' mechanism was the efficiency of the system in the presence of TBA. TBA is a known hydroxyl radical scavenger with a high reaction rate constant with hydroxyl radicals equal to 5×10^8 M^{-1}s^{-1} [20]. Figure 6 shows the efficiency of single and catalytic ozonation in the absence and presence of TBA, when zeolite-6 and talc-6 were used as catalysts. When TBA was added to the treatment system, the removal of p-CBA was 6.3%, 12.5% and 9.5% with the application of single ozonation or by the addition of zeolite and talk, respectively, after 30 min of oxidation time. The inhibition effect within the first 3 min of the relevant scavenging reaction was estimated at 94.5%, 99.3% and 99.5%, respectively. The presence of TBA also reduced the decomposition of ozone in all oxidation reactions (Figure S3). It is also worth noting that the application of higher ozone concentrations could not be beneficial in the relevant systems because ozone does not practically react with p-CBA ($k_{O3} < 0.15$ M^{-1}s^{-1}) [19].

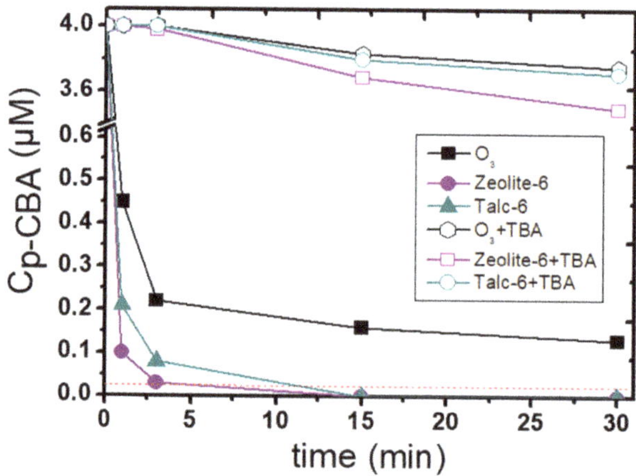

Figure 6. Effect of tert-butyl alcohol (TBA) in the removal of p-CBA during the application of single and catalytic ozonation processes; experimental conditions: $C_{p\text{-}CBA}$ 4 µM, C_{TBA} 0.4 mM, C_{O3} 2 mg/L, $C_{cat.}$ 0.5 g/L, pH 7, temperature 23 ± 2 °C.

3.3. Influence of the Main Experimental Parameters on the Removal of p-CBA

There are several factors that can affect the efficiency of the heterogeneous catalytic ozonation process. The most important among them are pH and temperature, which influence the stability of ozone in water [4]. The influence of experimental conditions was studied for the examined materials, especially in the cases of zeolite-6 and talc-6, calcined at 600 °C. The difference in the respective efficiencies between materials calcined at temperatures 600 °C and 800 °C was rather negligible. Also, the moderate calcination temperature of 600 °C is more practically feasible and less costly.

pH value is a very important aspect of the ozonation process because it highly influences the ozone decomposition and the subsequent production of hydroxyl radicals. As the pH raises, the decomposition of ozone accelerates (Figure S4). Figure 7 shows p-CBA removal when zeolite-6 and talc-6 were added to the system as catalysts for pH values 6, 7 and 8. As expected, higher pollutant degradation was observed at pH 8, independently of the catalyst that was added to the oxidation system. At pH 8, p-CBA was almost totally removed, already within the third min of oxidation time. In contrast, at pH 6 and for the same time (third min) the p-CBA removal was 91.7% and93% forzeolite-6 and talc-6, respectively. Less efficiency at this (slightly acidic) pH value could be attributed to the lower production of hydroxyl radicals, since p-CBA cannot practically react with ozone molecules, only hydroxyl radicals. Neutrally charged talc and slightly negatively charged zeolite were found to degrade p-CBA almost entirely after 15 min of reaction/oxidation time in pH range 6–8.

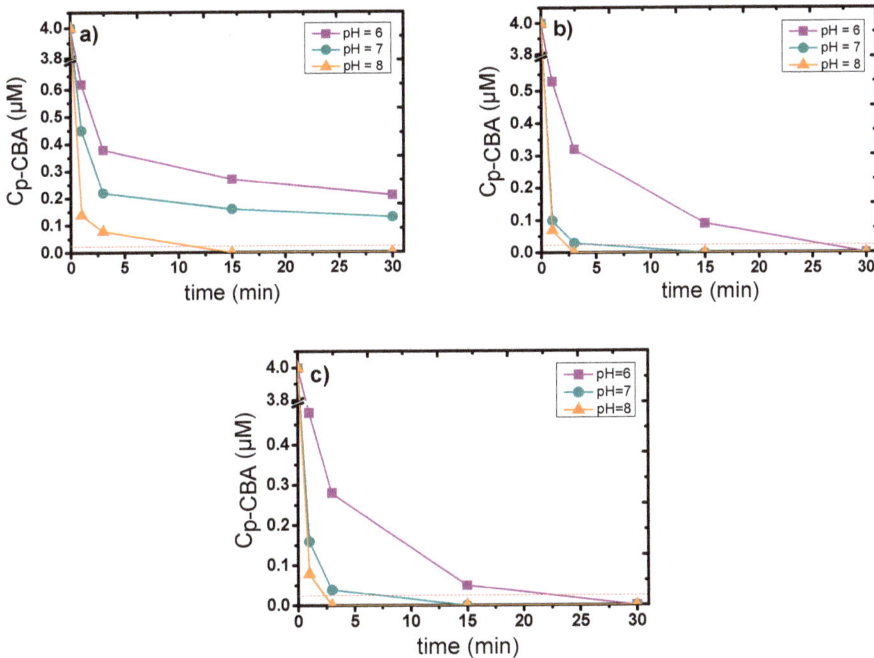

Figure 7. Influence of pH values on p-CBA removal by the application of (**a**) single ozonation and heterogeneous catalytic ozonation, using qualified catalysts:(**b**) Zeolite-6 and(**c**) Talc-6; experimental conditions: $C_{p\text{-}CBA}$ 4 μM, C_{O3} 2 mg/L, $C_{cat.}$ 0.5 g/L, temperature 23 ± 2 °C.

Furthermore, ozone molecules are more stable in an aqueous solution at lower temperatures; therefore, ozone decomposition reactions and the respective production of radicals are slower. For this reason, the removal efficiency of micropollutants was decreased by decreasing the solution temperature (Figure 8). At 15 °C, the production of hydroxyl radicals was limited, and the p-CBA was removed by 93.2% and 82% with the addition of zeolite-6 and talc-6, respectively, after 30 min of reaction/oxidation time. According to the van't Hoff rule, an increase in temperature of 10 °C will double the reaction kinetics rate. Although a rise in temperature can increase ozone decomposition and chemical reaction rates, it simultaneously decreases the dissolved ozone concentration in water [30]. Table 2 shows the escape of ozone in the gas phase at various temperatures under the assumption that the equilibrium of ozone in the gas/aqueous phase would be achieved quickly, resulting in lower ozone participation at the oxidation/reaction stage. These calculations are based on Wang et al.'s study [31] and are presented in the supplementary text S1. However, even in such an extreme theoretical hypothesis, ozone behavior depends on the temperature of the reaction. An increase in the temperature of an aqueous solution accelerates the escape of ozone in the gaseous phase (Table 2). Thus, increasing the reaction kinetic for p-CBA degradation by increasing the temperature from 15 °C to 25 °C exceeds the corresponding increase of ozone escape in the gaseous phase, while at 35 °C the higher escape of ozone in the gaseous phase results in lower degradation of p-CBA. Furthermore, increasing the temperature generally decreases the adsorption of micropollutants (Table S2) on the surface of catalysts, reducing the efficiency of heterogeneous catalytic ozonation reactions [30]. The actual ozone concentration during the experiments as a function of temperature is shown in supplementary Figure S5.

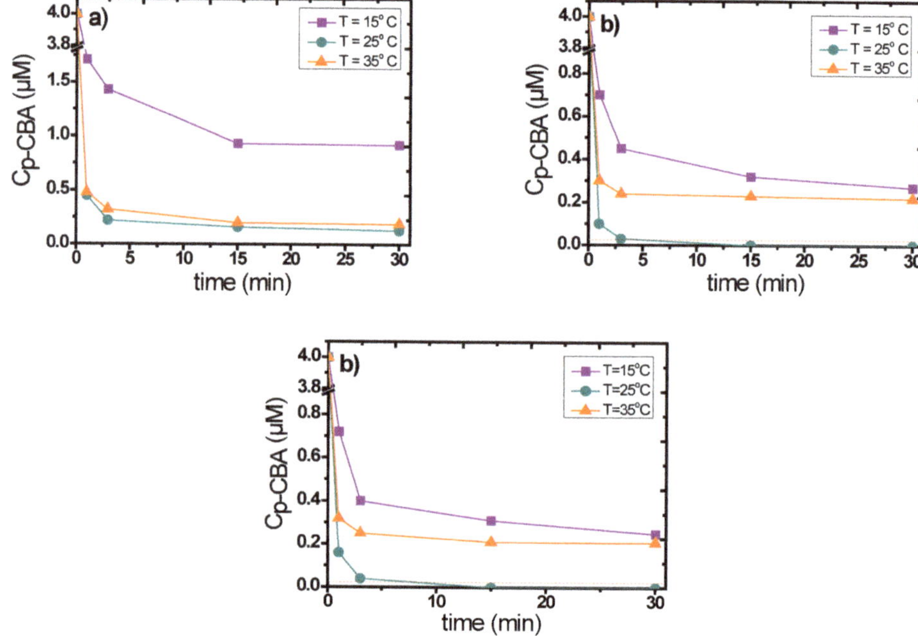

Figure 8. Influence of temperature on the removal of p-CBA by the application of (**a**) single ozonation and heterogeneous catalytic ozonation using optimum defined catalysts: (**b**) zeolite-6 and (**c**) Talc-6; experimental conditions: $C_{p\text{-CBA}}$ 4 μM, C_{O3} 2 mg/L, $C_{cat.}$ 0.5 g/L, pH 7.

Table 2. Theoretically calculated ozone escape from aqueous solution into air for various temperatures during the reaction/oxidation stage.

	Ozone Escape		
Reaction Time (Min)	**15 °C**	**25 °C**	**35 °C**
0.1	3.8%	4.7%	5.6%
1	5.8%	7%	8.3%
3	7.7%	9.4%	11.1%
15	9.6%	11.7%	13.9%
30	11.5%	14.1%	16.7%

In this study, the optimum temperature for the removal of pollutants by the application of heterogeneous catalytic ozonation was 25 °C. After 3 min of reaction time, the concentration of p-CBA was 34.4 μg/L, 4.7 μg/L and 6.3 μg/L in the O_3, zeolite-6/O_3 and talc-6/O_3 oxidation processes. When the temperature was raised to 35 °C, the removal efficiency for the same duration/oxidation time decreased by 31.3%, 87.5% and 68%, respectively. This reduction was mainly attributed to lower adsorption capacity (Table S2), as well as the increase of ozone escape towards the gas/air phase. Similar observations were also reported by Luo et al., 2018 [32].

4. Conclusions

In this study, the catalytic activity of three natural minerals, selected after preliminary screening experiments, was examined after thermal pre-treatment in the temperature range between 100° and 800 °C, during the application of heterogeneous catalytic ozonation process for the removal of p-CBA. The PZC of these solid materials was increased by thermal treatment. A PZC value around the pH value of the solution favors the efficient contact of ozone molecules with the catalyst surface and therefore can accelerate its decomposition, leading to the production of •OH radicals.

For the PZC values of the examined minerals approaching the circa-neutral solution pH (i.e., for the neutrally charged solids), the efficiency of the oxidation process increased. At pH 7 and 25 °C, the negatively charged kaolin surface showed no catalytic activity. Under the same conditions, talc (as a raw material with PZC 5.9) was found to present moderate catalytic activity, but after its calcination at 800 °C, performance increased by 56% even after the first min of oxidation time because the respective PZC value increased to 6.5. Furthermore, neutrally charged raw zeolite was found to efficiently remove p-CBA and increase the PZC value (to 7.2) after calcination further improved its performance.

The higher production of •OH and the subsequent higher efficiency of the oxidation process were found to be closely related to the neutrally charged surfaces of solid materials, acting as catalysts that can more effectively attract ozone molecules. The optimum conditions for all the examined parameters were a pH value of 8 and a temperature of 25 °C. Where the dissolution of ozone in the aqueous phase was satisfactory, the decomposition of ozone was fast and the production of •OH radicals was high. The removal of p-CBA was based on the mechanism of radical production and was verified by the addition of TBA, which substantially reduced the efficiency of the treatment system (in the range of 94.5% in the case of single ozonation).

Supplementary Materials: The following are available online at http://www.mdpi.com/2071-1050/12/24/10503/s1, Figure S1: (a) TG-DTA and (b) XRD diagram of kaolin; Peaks for K: Kaolin, I: Illite. Figure S2: Influence of the thermal pre-treatment on ozone decomposition in comparison to single ozonation: (a) zeolite, (b) talc, and (c) kaolin. Experimental conditions: ozone concentration 2 mg/L, catalyst concentration 0.5 g/L, p-CBA concentration 4 µM, pH 7 and temperature 23 ± 2 °C. Figure S3: Effect of TBA on ozone decomposition in the catalytic ozonation systems, as compared to singleozonation. Experimental conditions: ozone concentration 2 mg/L, catalyst concentration 0.5 g/L, p-CBA concentration 4 µM, TBA concentration 0.4 mM, pH 7 and temperature 23 ± 2 °C. Figure S4: Effect of pH value on the decomposition of ozone; (a) Single ozonation, (b) zeolite-6, and (c) talc-6. Experimental conditions: ozone concentration 2 mg/L, catalyst concentration 0.5 g/L, p-CBA concentration 4 µM and temperature 23 ± 2 °C. Figure S5: Effect of temperature on ozone decomposition; (a) Single ozonation, (b) zeolite-6, and (c) talc-6. Experimental conditions: ozone concentration 2 mg/L, catalyst concentration 0.5 g/L, p-CBA concentration 4 µM and pH 7. Table S1: Adsorption capacity of natural minerals thermally pre-treated at various temperatures for the case of p-CBA removal. Experimental conditions: catalyst concentration 0.5 g/L, p-CBA concentration 4 µM, pH 7, temperature 23 ± 2 °C and adsorption time 30 min. Table S2: Adsorption capacity of zeolite-6, and talc-6 for the case of p-CBA removal at various temperatures. Experimental conditions: catalyst concentration: 0.5 g/L, p-CBA concentration: 4 µM, pH: 7 and adsorption time 30 min. Text S1: Example of calculation, regarding the theoretical escape of ozone from the water during the experiment.

Author Contributions: Conceptualization, S.P., M.M. and A.Z.; methodology, S.P. and M.M.; validation, S.P., M.M. and A.Z.; formal analysis, S.P., E.K. and K.K.; investigation, S.P. and K.K.; resources, M.M. and A.Z.; data curation, S.P. and E.K.; writing—original draft preparation, S.P.; writing—review and editing, E.K., M.M. and A.Z.; visualization, S.P.; supervision, M.M. and A.Z.; project administration, A.Z.; funding acquisition, A.Z. All authors have read and agreed to the published version of the manuscript.

Funding: This research has been co-financed by the European Union and Greek national funds through the Operational Program Competitiveness, Entrepreneurship and Innovation, under the call RESEACH-CREATE-INNOVATE (project code: T1EDK-02397).

Conflicts of Interest: The authors declare no conflict of interest.

References

1. Cuerda-Correa, E.M.; Alexandre-Franco, M.F.; Fernández-González, C. Advanced Oxidation Processes for the Removal of Antibiotics from Water. An Overview. *Water* **2019**, *12*, 102. [CrossRef]
2. Stamm, C.; Räsänen, K.; Burdon, F.J.; Altermatt, F.; Jokela, J.; Joss, A.; Ackermann, M.; Eggen, R.I.L. Unravelling the Impacts of Micropollutants in Aquatic Ecosystems. In *Advances in Ecological Research*; Elsevier: Amsterdam, The Netherlands, 2016; Volume 55, pp. 183–223. [CrossRef]
3. Rogowska, J.; Cieszynska-Semenowicz, M.; Ratajczyk, W.; Wolska, L. Micropollutants in Treated Wastewater. *Ambio* **2020**, *49*, 487–503. [CrossRef] [PubMed]
4. Nawrocki, J.; Kasprzyk-Hordern, B. The Efficiency and Mechanisms of Catalytic Ozonation. *Appl. Catal. B Environ.* **2010**, *99*, 27–42. [CrossRef]
5. Gao, G.; Kang, J.; Shen, J.; Chen, Z.; Chu, W. Heterogeneous Catalytic Ozonation of Sulfamethoxazole in Aqueous Solution over Composite Iron–Manganese Silicate Oxide. *Ozone Sci. Eng.* **2017**, *39*, 24–32. [CrossRef]
6. Torres Sánchez, R.M.; Tavani, E.L. Temperature Effects on the Point of Zero Charge and Isoelectric Point of a Red Soil Rich in Kaolinite and Iron Minerals. *J. Therm. Anal.* **1994**, *41*, 1129–1139. [CrossRef]
7. Wang, B.; Xiong, X.; Ren, H.; Huang, Z. Preparation of MgO Nanocrystals and Catalytic Mechanism on Phenol Ozonation. *RSC Adv.* **2017**, *7*, 43464–43473. [CrossRef]
8. Zhu, H.; Ma, W.; Han, H.; Han, Y.; Ma, W. Catalytic Ozonation of Quinoline Using Nano-MgO: Efficacy, Pathways, Mechanisms and Its Application to Real Biologically Pretreated Coal Gasification Wastewater. *Chem. Eng. J.* **2017**, *327*, 91–99. [CrossRef]
9. Yuan, L.; Shen, J.; Chen, Z.; Liu, Y. Pumice-Catalyzed Ozonation Degradation of p-Chloronitrobenzene in Aqueous Solution. *Appl. Catal. B Environ.* **2012**, *117–118*, 414–419. [CrossRef]
10. Gao, G.; Shen, J.; Chu, W.; Chen, Z.; Yuan, L. Mechanism of Enhanced Diclofenac Mineralization by Catalytic Ozonation over Iron Silicate-Loaded Pumice. *Sep. Purif. Technol.* **2017**, *173*, 55–62. [CrossRef]
11. Yuan, L.; Shen, J.; Chen, Z.; Guan, X. Role of Fe/Pumice Composition and Structure in Promoting Ozonation Reactions. *Appl. Catal. B Environ.* **2016**, *180*, 707–714. [CrossRef]
12. Zhao, L.; Sun, Z.; Ma, J.; Liu, H. Enhancement Mechanism of Heterogeneous Catalytic Ozonation by Cordierite-Supported Copper for the Degradation of Nitro benzene in Aqueous Solution. *Environ. Sci. Technol.* **2009**, *43*, 2047–2053. [CrossRef] [PubMed]
13. Sui, M.; Sheng, L.; Lu, K.; Tian, F. FeOOH Catalytic Ozonation of Oxalic Acid and the Effect of Phosphate Binding on Its Catalytic Activity. *Appl. Catal. B Environ.* **2010**, *96*, 94–100. [CrossRef]
14. Kermani, M.; Kakavandi, B.; Farzadkia, M.; Esrafili, A.; Jokandan, S.F.; Shahsavani, A. Catalytic Ozonation of High Concentrations of Catecholover $TiO_2@Fe_3O_4$ Magnetic Core-Shell Nanocatalyst: Optimization, Toxicity and Degradation Pathway Studies. *J. Clean. Prod.* **2018**, *192*, 597–607. [CrossRef]
15. Liu, Z.-Q.; Tu, J.; Wang, Q.; Cui, Y.-H.; Zhang, L.; Wu, X.; Zhang, B.; Ma, J. Catalytic Ozonation of Diethyl Phthalate in Aqueous Solution Using Graphite Supported Zinc Oxide. *Sep. Purif. Technol.* **2018**, *200*, 51–58. [CrossRef]
16. Huang, Y.; Xu, W.; Hu, L.; Zeng, J.; He, C.; Tan, X.; He, Z.; Zhang, Q.; Shu, D. Combined Adsorption and Catalytic Ozonation for Removal of Endocrine Disrupting Compounds over $MWCNTs/Fe_3O_4$ Composites. *Catal. Today* **2017**, *297*, 143–150. [CrossRef]

17. Psaltou, S.; Karapatis, A.; Mitrakas, M.; Zouboulis, A. The Role of Metal Ions on P-CBA Degradation by Catalytic Ozonation. *J. Environ. Chem. Eng.* **2019**, *7*, 103324. [CrossRef]
18. Yi, H.; Zhao, Y.; Liu, Y.; Wang, W.; Song, S.; Liu, C.; Li, H.; Zhan, W.; Liu, X. A Novel Method for Surface Wettability Modification of Talc through Thermal Treatment. *Appl. Clay Sci.* **2019**, *176*, 21–28. [CrossRef]
19. Aljibouri, A.K.H.; Wu, J.; Upreti, S.R. Heterogeneous Catalytic Ozonation of Naphthenic Acids in Water. *Can. J. Chem. Eng.* **2019**, *97*, 67–73. [CrossRef]
20. Wang, Q.; Yang, Z.; Chai, B.; Cheng, S.; Lu, X.; Bai, X. Heterogeneous Catalytic Ozonation of Natural Organic Matter with Goethite, Cerium Oxide and Magnesium Oxide. *RSC Adv.* **2016**, *6*, 14730–14740. [CrossRef]
21. Clesceri, S.L.; Greenberg, E.A.; Trussell, R.R. *Inorganic Nonmetals, in Standard Methods for Examination of Water and Wastewater*, 17th ed.; American Public Health Association: Washington, DC, USA, 1989.
22. Bourikas, K.; Vakros, J.; Kordulis, C.; Lycourghiotis, A. Potentiometric Mass Titrations: Experimental and Theoretical Establishment of a New Technique for Determining the Point of Zero Charge (PZC) of Metal (Hydr) Oxides. *J. Phys. Chem. B* **2003**, *107*, 9441–9451. [CrossRef]
23. Perraki, T.; Kakali, G.; Kontori, E. Characterization and Pozzolanic Activity of Thermally Treated Zeolite. *J. Anal. Calorim.* **2005**, *82*, 109–113. [CrossRef]
24. Christidis, G. Chemical and Thermal Modification of Natural HEU-Type Zeolitic Materials from Armenia, Georgia and Greece. *Appl. Clay Sci.* **2003**, *24*, 79–91. [CrossRef]
25. Palenta, T.; Fuhrmann, S.; Greaves, G.N.; Schwieger, W.; Wondraczek, L. Thermal collapse and hierarchy of polymorphs in a faujasite-type zeolite and its analogous melt-quenched glass. *J. Chem. Phys.* **2015**, *142*, 084503. [CrossRef] [PubMed]
26. Parks, G.A. The Isoelectric Points of Solid Oxides, Solid Hydroxides, and Aqueous Hydroxo Complex Systems. *Chem. Rev.* **1965**, *65*, 177–198. [CrossRef]
27. Ardizzone, S.; Daghetti, A.; Franceschi, L.; Trasatti, S. The Point of Zero Charge of Hydrous RuO_2. *Colloids Surf.* **1989**, *35*, 85–96. [CrossRef]
28. Stankovic, J.; Milonjic, S.; Zec, S. The Influence of Chemical and Thermal Treatment on the Point of Zero Charge of Hydrous Zirconium Oxide. *J. Serb. Chem. Soc.* **2013**, *78*, 987–995. [CrossRef]
29. Kasprzyk-Hordern, B.; Dąbrowska, A.; Świetlik, J.; Nawrocki, J. Ozonation Enhancement with Nonpolar Bonded Alumina Phases. *Ozone Sci. Eng.* **2004**, *26*, 367–380. [CrossRef]
30. Gottschalk, C.; Libra, J.A.; Saupe, A. Ozonation of Water and Waste Water. In *A Practical Guide to Understanding Ozone and Its Application*; Wiley-VCH: Weinheim, Germany, 2010.
31. Wang, D.; Xu, H.; Ma, J.; Lu, X.; Qi, J.; Song, S. Strong Promoted Catalytic Ozonation of Atrazine at Low Temperature Using Tourmaline as Catalyst: Influencing Factors, Reaction Mechanisms and Pathways. *Chem. Eng. J.* **2018**, *354*, 113–125. [CrossRef]
32. Luo, L.; Zou, D.; Lu, D.; Yu, F.; Xin, B.; Ma, J. Study of Catalytic Ozonation for Tetracycline Hydrochloride Degradation in Water by Silicate Ore Supported Co_3O_4. *RSC Adv.* **2018**, *8*, 41109–41116. [CrossRef]

Publisher's Note: MDPI stays neutral with regard to jurisdictional claims in published maps and institutional affiliations.

 *sustainability*

Article

Loofah Sponges as Bio-Carriers in a Pilot-Scale Integrated Fixed-Film Activated Sludge System for Municipal Wastewater Treatment

Huyen T.T. Dang [1], Cuong V. Dinh [1], Khai M. Nguyen [2,*], Nga T.H. Tran [2], Thuy T. Pham [2] and Roberto M. Narbaitz [3]

[1] Faculty of Environmental Engineering, National University of Civil Engineering, No 55, Giai Phong Street, Hanoi 84024, Vietnam; huyendtt@nuce.edu.vn (H.T.T.D.); cuongdv@nuce.edu.vn (C.V.D.)

[2] Faculty of Environmental Sciences, VNU University of Science, 334 Nguyen Trai, Thanh Xuan, Hanoi 84024, Vietnam; tranthihuyennga@hus.edu.vn (N.T.H.T.); phamthithuy@hus.edu.vn (T.T.P.)

[3] Department of Civil Engineering, University of Ottawa, 161 Louis Pasteur Pvt., Ottawa, ON K1N 6N5, Canada; narbaitz@uottawa.ca

* Correspondence: khainm@hus.edu.vn

Received: 15 May 2020; Accepted: 5 June 2020; Published: 10 June 2020

Abstract: Fixed-film biofilm reactors are considered one of the most effective wastewater treatment processes, however, the cost of their plastic bio-carriers makes them less attractive for application in developing countries. This study evaluated loofah sponges, an eco-friendly renewable agricultural product, as bio-carriers in a pilot-scale integrated fixed-film activated sludge (IFAS) system for the treatment of municipal wastewater. Tests showed that pristine loofah sponges disintegrated within two weeks resulting in a decrease in the treatment efficiencies. Accordingly, loofah sponges were modified by coating them with $CaCO_3$ and polymer. IFAS pilot tests using the modified loofah sponges achieved 83% organic removal and 71% total nitrogen removal and met Vietnam's wastewater effluent discharge standards. The system achieved considerably high levels of nitrification and it was not limited by the loading rate or dissolved oxygen levels. Cell concentrations in the carriers were twenty to forty times higher than those within the aeration tank. Through 16S-rRNA sequencing, the major micro-organism types identified were *Kluyvera cryocrescens, Exiguobacterium indicum, Bacillus tropicus, Aeromonas hydrophila, Enterobacter cloacae, and Pseudomonas turukhanskensis.* This study demonstrated that although modified loofah sponges are effective renewable bio-carriers for municipal wastewater treatment, longer-term testing is recommended.

Keywords: integrated fixed-film activated sludge systems; modified loofah sponge; bio-carrier; microbial density; municipal wastewater

1. Introduction

One of the greatest challenges for developing countries is wastewater treatment. Rapid urbanization led to a more concentrated release of wastewaters into the environment, which has significantly impacted humans and wildlife. Organic compounds and nutrients from municipal or industrial wastewater contaminate ecosystems, e.g., rivers, ponds, or even groundwater [1]. Low-cost wastewater treatment utilizing local materials can contribute to addressing this challenge, especially in developing countries. Wastewater stabilization ponds or constructed wetlands are known as appropriate low-cost wastewater treatment technologies, but they have large land area requirements, and the treatment efficiency is rather difficult to control. In developing countries, the conventional activated sludge process (CAS) has been the standard treatment method. During the last decades, new fixed-film processes, such as moving bed biofilm reactors (MBBR) and integrated fixed-film activated sludge (IFAS) systems, have also been widely applied [2,3]. These processes have performed well in terms of organics and nitrogen removal in both municipal [4,5] and industrial wastewater treatment applications [6]. The key differences between MBBR and IFAS are the possible longer hydraulic retention time and recycle of sludge in the IFAS system [7].

IFAS is a biological wastewater treatment process, which combines suspended growth and attached growth processes by adding free-floating biofilm carriers into the aeration tank [2,8]. Biofilm carriers, which have a lower density than water, are kept in suspension, moving throughout the tank volume by the aeration. The relatively large surface of the carriers serves as physical support and protection for attached microbiological growth and leads to large bacterial populations throughout the reactor. A key advantage of IFAS over CAS is that the biomass attached to the bio-carriers is retained within the system. This leads to higher solids retention times and biomass concentrations, thus reducing the required reactor volume/footprint [9]. Conversion of CAS systems to IFAS systems is relatively inexpensive and it will yield higher system capacities or improved removal efficiencies [7]. IFAS also produces better settling solids, so the secondary clarifier can operate at higher solid loadings and produce lower concentration effluents [9]. The attached biomass also results in more stable nitrification [7].

The most common biofilm carriers used in IFAS and MBBR systems are polyethylene bio-carries [10] and polyurethane sponges [11]. However, the cost of plastic bio-carriers makes them less attractive for applications in developing countries. In addition, plastic carriers are environmentally unfriendly materials as they may generate secondary waste, i.e., plastic waste. The application of natural materials as carriers for MBBRs or IFASs has attracted significant attention [12]. Examples of alternative materials tested as bio-carriers include black volcanic ashes [13], natural zeolites [14], fungal pellets [15], diatomaceous earth [16], and Moringa oleifera seeds [17].

Loofah is an annual herbaceous plant from the cucurbitaceous family, and its fully developed fruit is the source of the loofah scrubbing sponges normally used in bathrooms and kitchens. They are also used for other applications such as for reinforcing composite materials [18]. Loofah sponges are lightweight, highly porous and have a large surface area, which makes them suitable as a bio-carrier material in IFAS and MBBR systems. So far, the most common research interest for loofah sponges is their adsorption capability, e.g., adsorption of phenol [19], heavy metal [20], and organic compound [21]. Loofah sponges are composed of hemicellulose (22%), α-cellulose (60%), lignin (10.6%), and others (7.4%) [19]. These components are biodegradable in water [22,23]. In order to enhance the durability of loofah sponges in aqueous solutions, some studies have modified them. Hideno et al. [24] modified the loofah sponges by immersing them in an acetic anhydride solution to create acetylated sponges for the immobilization of cellulase-producing microorganisms. On the other hand, they soaked the luffa sponges in a $Ca(OH)_2$ solution to increase the durability of the sponges to up to 10 days in a lab-scale bioreactor [25]. It should be noted that the modification with the alkaline solution was reported to reduce the leaching of hemicelluloses, waxes, impurities, and lignin from the fibers [26].

The objective of this study is to modify loofah sponges and use them as bio-carriers in a pilot-scale IFAS unit for the treatment of municipal wastewater. The experiments evaluated the organic compound and nitrogen removals, as well as the bacterial density and bacterial community composition. This study is novel in that it presents the first pilot-scale IFAS wastewater treatment study using loofah sponges as bio-carriers. This treatment approach increases the sustainability of IFAS by using a low-cost renewable material as bio-carriers.

2. Materials and Methods

2.1. Preparation and Modification of Loofah Sponges

Loofah sponges were collected and cut into pieces (about 5-6 cm long). The sponges were rinsed with distilled water (DI) and oven-dried at 105 °C for 24 h. The objective of the loofah sponge modification was to increase their durability. To strengthen and increase the fiber hardness and service life, Liu et al. (2016) coated their loofah sponges with a $CaCO_3$ layer by soaking them in a lime solution. The $CaCO_3$ coating was formed based on the following equation:

$$Ca(OH)_2 + CO_2 \rightarrow CaCO_3\downarrow + H_2O \tag{1}$$

In this study, the modification process was carried out as follows. First, pieces of loofah sponges were soaked in 500 mL of a 30% $Ca(OH)_2$ solution (ACS reagent $\geq$95.0%, Sigma-Aldrich) for 48 h, followed by drying at room temperature for 24 h. Second, the samples were then immersed in 200 mL of 30% by volume commercial Acrylic - Styrene polymer solution (Bondex, India) for 1 h. The polymer solution had a viscosity of 3–8 (kg m^{-1} s^{-1}), at 30 °C and pH = 8–10, and is alkali resistant. The polymer coating was intended to increase the loofah fiber's water tolerance and possibly increase its surface area. The polymer solution was received as an emulsion which was converted into a suspension (via dispersion) by mixing it with a small amount of water. The main benefits of polymer coating were its plasticization and the possibility of greater microorganism adhesion. Third, the samples were removed from the polymer solution, allowed to drain and dry at room temperature for 24 h, and then stored for the experiments. The pristine and modified loofah samples were characterized before conducting the tests.

To assess the impact of different coating strategies and the durability of these modified sponges in water, a set of wastewater contact experiments were conducted using five types of loofah sponge samples, including a pristine loofah sponge (Sample 1), a $CaCO_3$-coated loofah sponge (Sample 2), a loofah sponge sample coated by immersion in both a $Ca(OH)_2$ solution and a 30% polymer solution (Sample 3), a loofah sponge sample coated by spraying with the polymer solution (Sample 4), and a loofah sponge sample coated by immersing it in 30% polymer solution only (Sample 5). This test was implemented prior to conducting the IFAS experiment to evaluate the organic and nitrogen removal efficiencies.

2.2. IFAS Pilot-Scale Experiment

The pilot-scale IFAS system was set up at Kim Lien wastewater treatment plant (WWTP) in downtown Hanoi (Figure 1). This system was designed for the simultaneous removal of organic compounds and nitrogen compounds. The system consists of an anoxic compartment, a larger aerobic IFAS bioreactor compartment, and a sedimentation compartment. The feed wastewater was the Kim Lien wastewater treatment plant's primary effluent. It was pumped to the anoxic compartment (Figure 1), where a mixer (2) was employed to ensure that the solids did not settle. The wastewater then passed to the aerobic compartment which contained the loofah sponges, which were moving within the tank due to the aeration. From there, the wastewater flowed into the sedimentation chamber. The sedimentation chamber effluent was stored in a separate water tank. As in Figure 1, the blower (1) was used to supply air for the bioreactor. In addition, wastewater was recycled from the aerobic compartment to the anoxic compartment by a pump (3) and sludge was recycled from the sedimentation compartment to the anoxic compartment by a pump (4) to enhance the nitrogen removal by denitrification. The tank dimensions and operational parameters are presented in Table 1. The pilot system flowrate was 1 m^3 d^{-1}. Note that the hydraulic retention time (HRT) of the aeration/bioreactor compartment was within the 3 to 7 h conventional range [7]. The anoxic zone's HRT was purposely chosen to be slightly longer than normal (1–2 h), this was to ensure denitrification is not hindered. The pristine and modified loofah sponges were put into the MBBR tank at a bulk volume filling ratio of 30%.

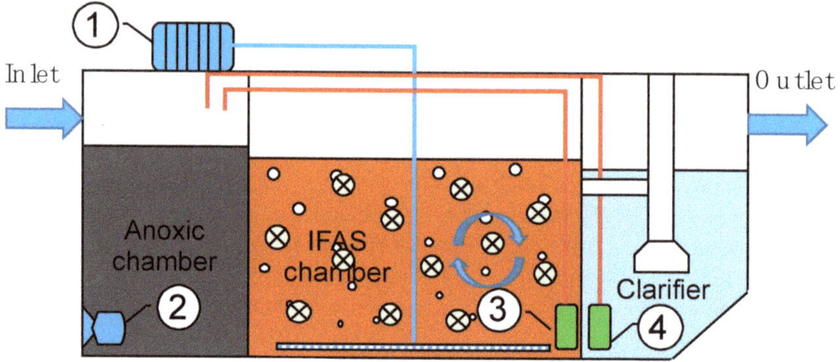

Figure 1. Experimental setup of the integrated fixed-film activated sludge (IFAS) system.

Table 1. Pilot-scale moving bed biofilm reactors (MBBR) specifications and operational conditions.

Treatment Compartment	Parameters	Value
Anoxic compartment	Wet dimensions, $L \times W \times H$ (m)	$0.4 \times 0.4 \times 0.7$
	Water volume (m^3)	0.112
	Hydraulic retention time (h)	2.65
Aerobic compartment (MBBR)	Wet dimensions $L \times W \times H$ (m)	$0.4 \times 0.8 \times 0.7$
	Hydraulic retention time (h)	5.3
	Water volume (m^3)	0.224
	Circulation rate, R	$0.5Q$
	Carrier filling rate	30%V
	Air supply ($m^3 \ min^{-3}$)	0.1
Sedimentation compartment	Wet dimensions, $L \times W \times H$ (m)	$0.4 \times 0.4 \times 0.7$
	Sludge circulation rate, R_b	$0.5Q$

During the initial experiments with pristine loofah sponges, the feed wastewater had a pH of 7.4 ± 0.25, a temperature of 25 ± 4.5 °C, a chemical oxygen demand (COD) concentration of 175.2 ± 32.5 mg L^{-1}, and total nitrogen (TN) concentration of 45.7 ± 1.9 mg L^{-1}. The ammonia and COD concentrations fluctuated due to dilution with the occasional rainfall events, as the Kim Lien WWTP is served by a combined sewer system. In the aerobic compartment, the air was provided at a rate of $0.1 \ m^3 \ min^{-1}$ to ensure the attached biomass had sufficient dissolved oxygen and the bio-carriers materials remained suspended during the reaction process. Because of logistic limitations, the duration of the experiments was three months. The IFAS run with the unmodified loofah fibers carriers was conducted first, and it was followed by the IFAS run using the modified loofah fiber carriers. The pilot-scale system was initially seeded with activated sludge which was taken from the secondary settling tank of the Kim Lien WWTP. During the tests, samples were collected from the inlet, the MBBR compartment and the sedimentation compartment, and then analyzed for COD, TN, N-NH$_4$, dissolved oxygen (DO), pH and temperature. The samples were collected and analyzed daily for pH, temperature, and dissolved oxygen (DO), and twice per week for the other parameters (TN, N-NH$_4$, COD).

2.3. Analytical Method

The analysis of the above water quality parameters was performed according to the procedures in "standard methods for the examination of water and wastewater" [27].

The structure of loofah was evaluated by an optical microscope (Carl Zeiss, Germany). The surface area was analyzed by the Brunauer-Emmett-Teller (BET) method using a Gemini VII 2390 V1.02T analyzer (Micromeritics Instrument Corp., Norcross, GA, USA).

The bio-carrier specific volumetric filling ratio (V_{FR}) was determined by filling an empty 1-L beaker with loofah sponges, adding water to the 1-L mark, then removing the sponges and measuring the volume of water (V_2). Due to the occupancy of carriers, the volume of water remained in the beaker (V_2) shall be less 1 L. V_{FR} is defined by the following equation:

$$V_{FR} = \frac{V_2}{V_1} \times 100\% \tag{2}$$

where V_1 is the volume of water in a 1 L beaker (V_1 = 1 L), and V_2 is the volume of water remained in a 1 L beaker after removing the loofah sponges (L).

The density of loofah sponges was calculated using the following equation:

$$\rho = \frac{m}{V} \tag{3}$$

where m (g) and V (cm^3) are the mass and volume of the loofah sponge.

The microbial population and density were determined by colony culture and counting methods. The microorganisms in loofah samples were cultured in Luria-Bertani (LB) medium including peptone (15 g L^{-1}), agar (15 g L^{-1}), yeast (5 g L^{-1}) and NaCl (5 g L^{-1}). After two days, the colonies were counted to estimate the density as Colony Forming Units (CFUs) (Vietnam standard method TCVN 9716:2013, or ISO 8199:2005). In addition, 16S-rRNA analysis was performed three times during the testing, to determine the phylogenetic information of the isolated microorganism. The procedure is described in detail elsewhere [28,29]. Microbial analysis was performed for the following samples: mixed liquor from the aeration tank of the Kim Lien WWTP (as a reference); the solution in the pilot-scale aeration compartment; and the bio carriers within the IFAS pilot system.

The Spearman correlation coefficients were calculated for pairwise comparisons of organic loading rate and COD removal, and the evaluation of the correlation between ammonium loading rate and its reduction during the nitrification process. This was performed using XLSTAT 2019.3.1 (Addinsoft, USA).

3. Results and Discussion

3.1. Characterization of Loofah Sponges before and after Modification

Figure 2a, b show images of the pristine and modified loofah sponge bio-carriers. It can be seen that the sponges have a porous and fibrous structure, which can provide a favorable environment for the attached bacterial growth. Additionally, the modified loofah sponge pieces are covered by a white layer. This layer was created by the $CaCO_3$ precipitate and polymer modifications of the loofah fibers (Figure 2b). Figure 2c,d presented the optical microscope images of two carrier types. Notably, after the $CaCO_3$ precipitate and polymer modifications, a thin layer is observed on the fiber structure (Figure 2d). The precipitate and film layer could represent a film that prevents water from penetrating into the loofah fibers and, thus, enhances the durability of the loofah fiber bio-carriers. The images also show that the $CaCO_3$ and polymer layer did not fill the sponge's voids, the modified sponges maintained the original mesh-like structure.

Figure 2. (**a,b**) Real images and (**c,d**) microscope images (×10 magnification) of the pristine and modified loofah sponges, respectively.

Table 2 shows the key characteristics of pristine and modified loofah sponges. As shown, the densities of the pristine loofah sponge (0.25 g cm^{-3}) and the modified loofah sponge (0.26 g cm^{-3}) samples were significantly less than 1 g cm^{-3}, indicating that the loofah fiber carriers will float in water, and be suitable as bio-carriers. The slight increase in the density of the modified loofah sponge is likely due to the formation of the CaCO$_3$ precipitate and polymer layer on the material surface. The average carriers' filling ratios were relatively similar, that were 87.5% and 82.5% for the pristine and modified samples, respectively. This confirms that the CaCO$_3$-polymer modification did not significantly alter the porous structure of the loofah sponges. Moreover, the value of the BET specific surface area of the modified loofah sponge samples was 0.875 m^2 g^{-1}, which was higher than that of the pristine loofah samples of 0.018 m^2 g^{-1}. The higher surface area is attributed to the successful CaCO$_3$/polymer-composite coating of the loofah sponge surface, in which the higher surface area is attributed primarily to micropores in the coating layer. As discussed by Yuan et al. [30], a larger specific surface area would normally result in higher biomass attachment. Overall, the chemical modification of loofah sponges coated the framework structure and accordingly should increase the durability.

Table 2. Characteristics of the pristine and modified loofah sponges.

Parameters	Pristine Samples	Modified Samples
Material Size (mm)	30–50	30–50
Density (g cm^{-3})	0.25	0.26
Specific Carrier Filling ratio (%)	87.5% ± 3%	82.5% ± 3%
BET surface area (m^2 g^{-1})	0.018	0.875

3.2. Evaluation of Durability of Loofah Sponges in Water

The durability test of the materials was performed by soaking the pristine loofah sponge and several differently modified loofah samples into the active bioreactor tank (Figure 3a). This is a realistic approach, as the loofah sponge carriers were exposed to the real operating environment of a municipal wastewater treatment plant. It was found that after 5 days due to decomposition, only a portion of the pristine loofah sponge (sample 1) remained (Figure 3b). The samples coated by a layer of $CaCO_3$ or polymer (samples 2, 4, and 5) degraded within three weeks. These results are consistent with the findings of Do et al. [25], whose loofah sponges coated with $CaCO_3$ only lasted up to 10 days in the lab-scale MBBR tank. Only sample 3, coated by both $CaCO_3$ and polymer, remained intact after 30 days of operation (Figure 3c). Thus, the protective film of $CaCO_3$ precipitate and polymer successfully prevented the destruction of the loofah sponge during the contact with the wastewater and microorganisms. The combination of $CaCO_3$ and polymer coating increased the loofah sponge's water durability beyond that provided by $CaCO_3$ coating alone. These modified sponges, coated with both $CaCO_3$ and polymer, were selected for a long-term test in the IFAS system to evaluate the organic and nutrient removal.

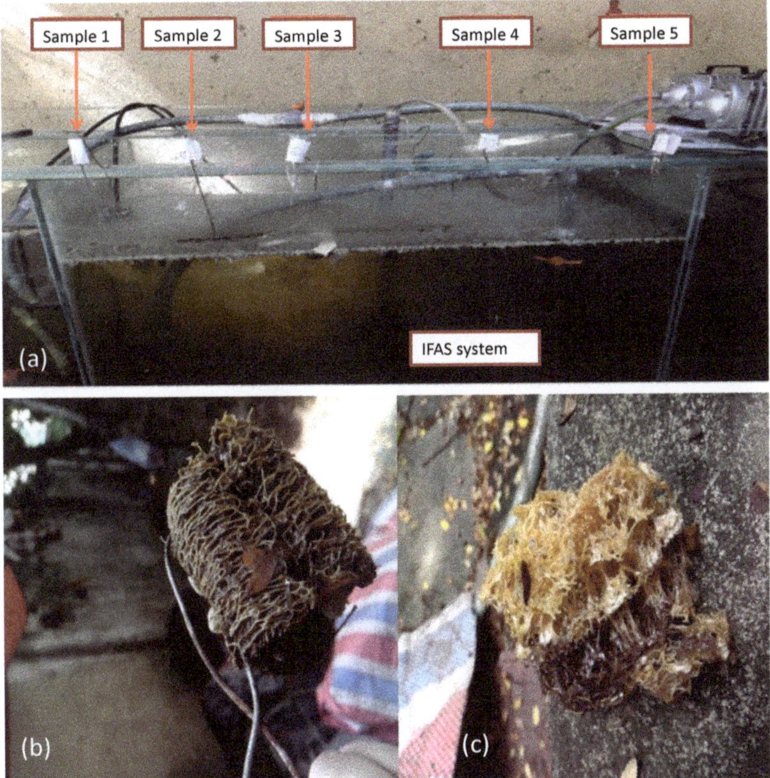

Figure 3. Evaluation of durability of loofah sponges in IFAS tank (**a**) the arrangement of 5 samples; (**b**) the sample 1 after 5 days; (**c**) the sample 3 after 30 days.

3.3. IFAS Removal of Organic Compounds

The COD removal of the IFAS system using the pristine loofah sponges as bio-carriers is presented in Figure 4a. The COD of the influent wastewater was 175.2±32.5 mg L^{-1} (Figure 4a), so this was a relatively low organic strength municipal wastewater. The COD of the effluent was 35.2 ± 18.9 mg L^{-1}, which is lower than Vietnam's COD standard for treated wastewater of 75 mg L^{-1} [31] (Figure 4a). The IFAS/pristine loofah carriers' system initially achieved a high average COD removal (79% ± 13%). However, the COD removal decreased to 62% after 15 days. It is speculated that this was caused by the structural degradation of the loofah sponges, which decreased the mass of attached microorganisms that were retained within the bioreactor.

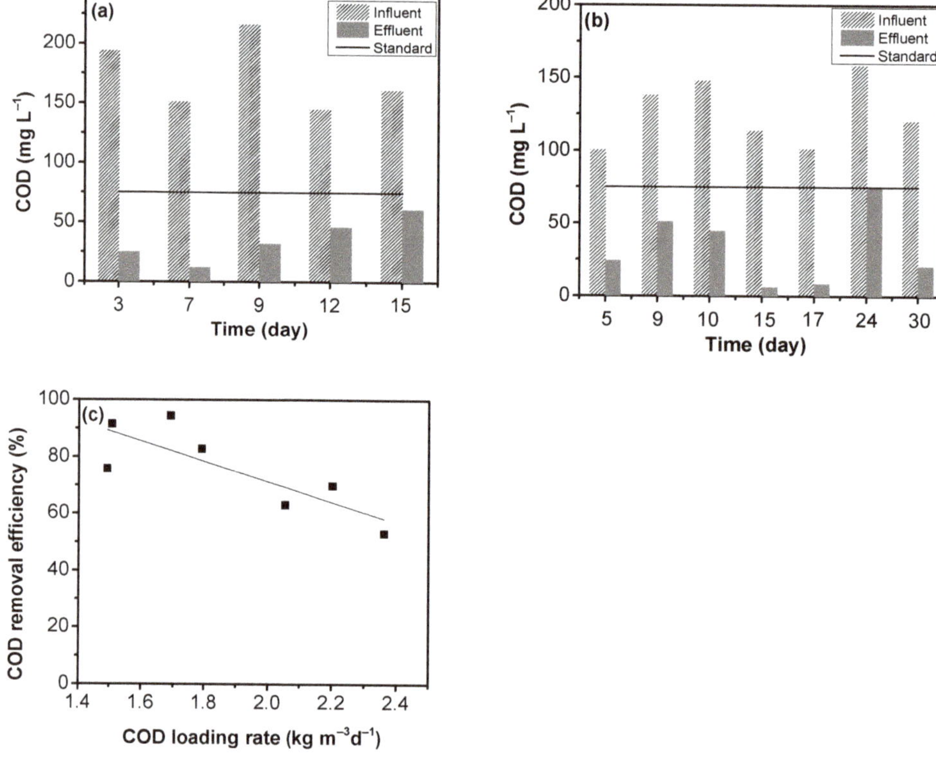

Figure 4. COD removal of the (**a**) pristine and (**b**) modified loofah sponges and (**c**) correlation of COD removal and COD loading rate in the IFAS/modified loofah sponges' system.

Figure 4b presents the IFAS/modified loofah sponge system's influent and effluent COD concentrations for the 30-day run. It should be noted that by the second week, the COD removals exceeded 90%, however, on the 24th day, there was a significant drop in the COD removal. This was attributed to the degradation of some modified loofah sponge bio-carriers, so some new modified loofah sponges were added to replace the degraded ones, and this improved the COD removals (Figure 4b). And by the 30th day, the system achieved 82.8% COD removal. Given that the feed wastewater was real municipal wastewater, the research group had no control over the feed concentrations and composition. The IFAS/modified loofah bio-carrier run had lower feed COD concentrations than the IFAS/pristine loofah bio-carrier run (125.8 ± 22.9 mg L^{-1} versus 175.2 ± 32.5 mg L^{-1}). Thus, the former had lower COD loadings that likely explains the higher COD removals for this run. In addition, the organic compounds in the feed wastewater also included a non-biodegradable fraction, so the degradable COD removals were even higher than the percentages reported above.

Overall, the modified loofah sponge carriers can help the IFAS system remain effective for a longer time than the pristine loofah sponge carriers. In addition, the percent COD removals efficiency of the IFAS using loofah bio-carrier is comparable with commercial plastic carriers (40–95%) [32,33].

Figure 4c presents the relationship between the COD fraction removed and the COD loading rate (kg COD m^{-3} day^{-1}) in the IFAS/modified loofah sponge pilot. This loading rate is based on the volume occupied by the sponges, i.e., 30% of the total. Statistical analysis showed a negative correlation (the Spearman coefficient $= -0.71$, $p < 0.05$), and a linear fit of data also presented a negative correlation with $R^2 = 0.662$. This indicates that the COD removed fraction decreased with the increasing COD loading rate from 1.4 to 2.4 kg m^{-3} d^{-1}. This is consistent with previous studies [34,35]. In fact, the COD removal performance may depend on a variety of factors such as the bacterial density, the bacteria types, and the operating conditions (i.e., DO, pH, and temperature) [33].

3.4. Removal of Nitrogen Compounds

The nitrogen removal performance of the IFAS/pristine and modified loofah sponge pilots are shown in Figure 5. For both runs, there was a significant fluctuation in the influent TN concentration. During the IFAS run with the pristine sponges, the influent TN concentrations were higher than during the IFAS- modified bio-carrier run (45.7 ± 1.9 mg L^{-1} versus 29.7 ± 9.3 mg L^{-1}). As observed, after 9 days the IFAS/pristine loofah sponge system achieved a relatively high TN removal (i.e., 80.3%) (see Figure 5a), however, the nitrogen removal gradually decreased after 15 days (i.e., 47.4%). This is likely due to the degradation of the loofah fibers after 15 days, as discussed above. The influent and effluent TN concentrations for the IFAS/modified loofah sponge bio-carriers are presented in Figure 5b. In all the testing, the IFAS/modified loofah bio-carrier system's effluent met Vietnam's 20 mg TN L^{-1} standard [31]. It should be noted that the TN removals also decreased on the 24th day of the run, however, the addition of new bio-carriers did not result in recovery as for the COD removals. This may be due to the fact that nitrifying bacteria grow more slowly than heterotrophs, and thus require more time to recover. The average percent TN removal using the modified carriers were superior to those observed for the pristine loofah bio-carriers, $71 \pm 8\%$ versus $53 \pm 22\%$. This may have been influenced by the lower feed TN concentrations during the IFAS/modified loofah bio-carrier run.

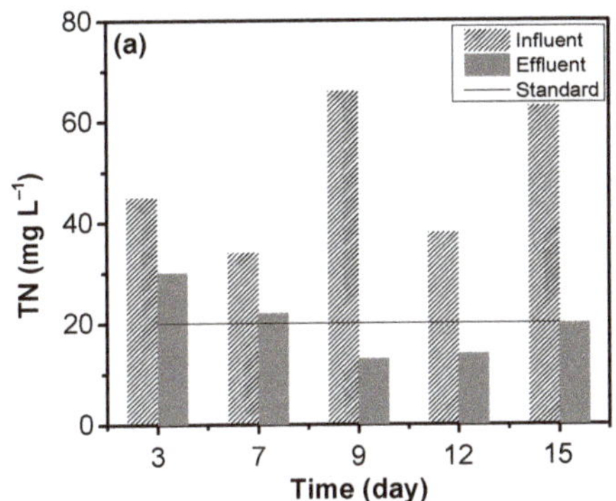

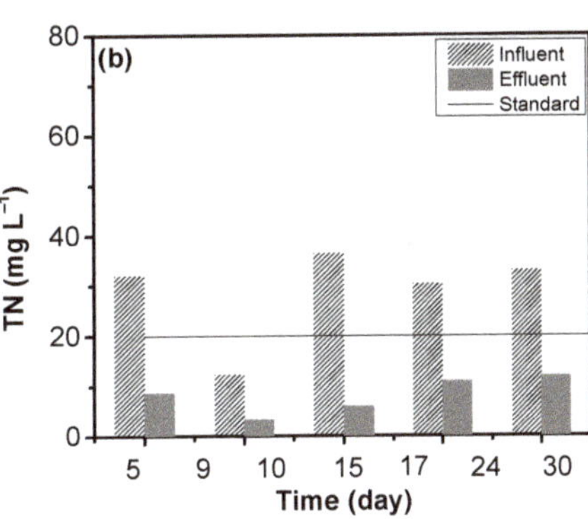

Figure 5. Nitrogen removal of the IFAS pilot using the (**a**) pristine and (**b**) modified loofah sponges.

The IFAS/modified loofah sponge run also investigated ammonia removals. TN is composed of ammonia, organic nitrogen, nitrite, and nitrate. The feed concentrations during this run also fluctuated, the TN concentration was 29.7 ± 9.3 mg N L^{-1} and the NH$_4$ was 19.7 ± 6.9 mg N L^{-1}. Given that raw wastewater generally have virtually no NO$_2$ and NO$_3$, the raw wastewater contained approximately 10 mg L^{-1} in the form of organic nitrogen. The relationship between nitrification rate and ammonium loading rate was investigated on the IFAS/modified loofah sponge run. (Figure 6a). The wide range in the ammonia loadings was due to the fairly wide range of ammonia feed concentrations. It was found that the data in Figure 6a fit a straight line rather well ($R^2 = 0.995$). Further statistical analysis confirmed the positive correlation with the Spearman coefficient of 0.997, $p < 0.05$. The slope of this line represents the percent ammonia removal, and the constant slope shows that the IFAS/modified loofah bio-carrier system was very consistent. The percent NH$_4$-N removal was $90.4 \pm 4.5\%$, so nearly complete nitrification was achieved. This high level of nitrification is logical given the low organic loadings applied to the system, i.e., the nitrifying bacteria were not significantly impacted by competition with the heterotrophs that oxidize the COD. The straight-line results in Figure 6a also demonstrate that the ammonia loading rate does not limit ammonia removal. Figure 6b shows the impact of DO on the fraction of NH$_4$-N removed. It shows that DO levels in the range of 2–5 mg L^{-1} do not have a significant impact on the level of nitrification. Thus, it can be concluded that the IFAS system with CaCO$_3$/polymer modified loofah sponge bio-carriers can successfully treat the organic and ammonium of municipal wastewater. The loofah sponge modification was critical in improving the durability of the eco-friendly carriers. Longer-term testing is recommended to evaluate their long-term durability.

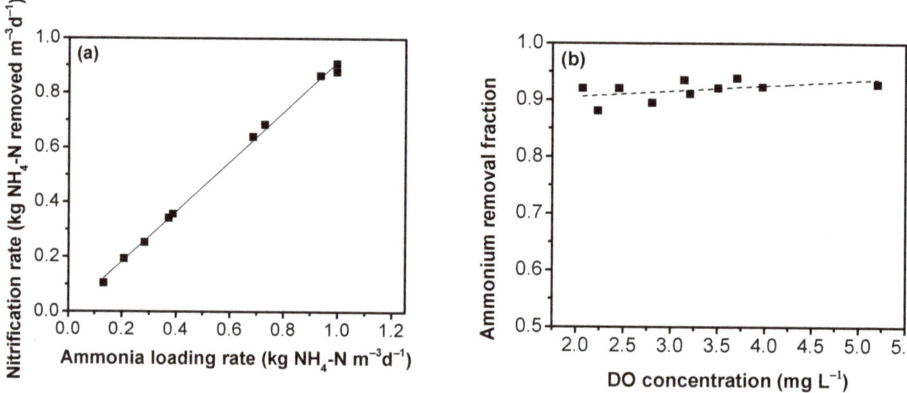

Figure 6. Nitrification analysis: (**a**) nitrification rate and ammonium loading rate and (**b**) DO and ammonium removal rate.

It is noteworthy that the IFAS/modified loofah bio-carrier system was capable of removing approximately 90% of the ammonia but only 71% of the TN. This may be because part of the TN is composed of organic nitrogen compounds, some of which may not be degradable [7]. Another possibility is that denitrification in the anoxic compartment was not fully effective due to difficulties in maintaining anoxic conditions, because of the large rate of mixed liquor recycle to the anoxic tank. At least on one occasion, the DO levels increased above 0.5 mg/L.

3.5. Microbial Evaluation

The density of microorganisms and related types of microorganisms were evaluated during the second IFAS phase with the modified bio-carriers. In Table 3, day "0" indicates the time when the microorganisms were fully adapted to the influent wastewater and grew well in the reactor. On that day, the microorganism density in the oxic tank of Kim Lien WWTP was 2.6×10^5 CFU mL^{-1}, while it was 1.5×10^4 CFU mL^{-1} in the IFAS tank solution, and an even higher bacterial density in the bio-carriers (6.7×10^5 CFU gr^{-1}). On day "0", the modified sponges had already been in the IFAS system for one week and their bacterial density was higher than the control modified loofah sponges (5.33×10^5 CFU gr^{-1}). So this was evidence that there was microorganism growth on the sponges once they were put in the system. As the system continued its operation, the density of microorganisms increased proportionally. This is logical because there was bio-growth within the tanks and on the fibers (carriers). Most of the time, the accumulation of microorganisms was higher in the IFAS tank than the oxic tank of the full-scale plant. This showed that the existence of bio-carriers helped increase the biological community and activity in the aerated reactor. In addition, the microbial growth was significant on bio-carriers (twenty to forty times higher in loofah sponges than in IFAS tank solution). The increasing biofilm formation can also be seen when comparing the color in active loofah sponges and the clean ones (Figures 2 and 3). On the 24th day, the bacterial mass on the carriers is much lower. The drop in bacterial mass on the carriers was much more significant than expected based on the visual observations of the modified loofah bio-carrier degradation. So presumably the sponges degraded to the point that they could not support as large a biofilm, or possibly the biofilm was using components of the sponge as substrate and they exhausted the supply of these components.

Table 3. The density of microorganisms in the IFAS/modified loofah sponges' system.

Operational Time (Day)	In Oxic Tank of Kim Lien Wwtp (CFU mL^{-1})	In Ifas with Modified Loofah Sponges (CFU mL^{-1})	On Modified Loofah Sponges within the IFAS (CFU gr^{-1})	On Pristine Modified Loofah Sponges (CFU gr^{-1})
0	2.6×10^5	1.6×10^4	6.7×10^5	
2	5.2×10^5	5.0×10^4	16.9×10^5	
6	6.7×10^5	25.5×10^4	187.0×10^5	5.33×10^5
12	14.1×10^5	129.0×10^4	290.0×10^5	
24	22.8×10^5	3.92×10^4	11.3×10^5	

The treatment efficiency in terms of COD and TN removal also decreased at this time. It confirms the fact that microorganisms played a critical role in transforming organic and ammonia in the reactor. It should be noted that after the 24th day, some new modified loofah sponges were added to replace the degraded ones, so the COD and TN removals increased again (Figures 4 and 5). Based on these results, for better treatment efficiency one should regularly replace the modified loofah sponges as their working life span is approximately four weeks.

Understanding the dominant microorganisms in suspension and the bio-carriers would help optimize conditions for their growth and maximize pollutant removal. The microorganism strains developed in the tank system and on modified loofah sponges were collected on days "12" and "24" for the microbial analysis to determine the dominant microorganisms. Results from dendrograms were calculated with the un-weighted pair group method and combined with arithmetic mean algorithm (UPGMA) clustering (Pearson correlation coefficient) of 16S-rRNA patterns. They identified six main groups of microorganism strains: (1) Common strain: round colonies, thick, diameter between 1–2 mm, white, most abundant (below Figure 7a, labeled 5.3C1), (2) Branched strain: round colonies, branched at the edge, diameter between 3–4 mm, white (labeled 5.2MS2) (Figure 7a), (3) Red strain: round colonies, thick, red, more intense color in the center, diameter between 2–3 mm (labeled 5.2PR3) (Figure 7a), (4) Orange strain: round colonies, orange, diameter between 1–2 mm (labeled 5.2MS4), (5) Yellow strain: round colonies, yellow, diameter between 1–2 mm (labeled 5.1MR5) (Figure 7a) and (6) White strain: curve-round colonies, smooth surface, diameter 4–15mm (4C1, 4C2, 4C4) (Figure 7b).

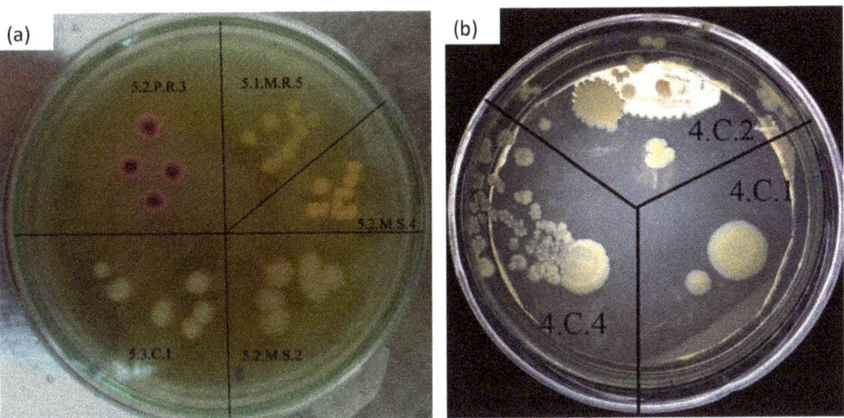

Figure 7. Clusters of isolated strains based on the 16S rRNA gene sequence analysis, with (**a**) Common strain (5.3C1), Branched strain (5.2MS2), Red strain (5.2PR3), Orange strain (5.2MS4), Yellow strain (5.1MR5), (**b**) White strain (4C1, 4C2, 4C4).

Based on the 16S rRNA gene sequence analysis of these samples, some major microorganism strains commonly found in wastewater were identified (Table 4). They include:

10. Liu, J.; Zhou, J.; Xu, N.; He, A.; Xin, F.; Ma, J.; Fang, Y.; Zhang, W.; Liu, S.; Jiang, M. Performance evaluation of a lab-scale moving bed biofilm reactor (MBBR) using polyethylene as support material in the treatment of wastewater contaminated with terephthalic acid. *Chemosphere* **2019**, *227*, 117–123. [CrossRef] [PubMed]

11. Zhang, X.; Chen, X.; Zhang, C.; Wen, H.; Guo, W.; Ngo, H.H. Effect of filling fraction on the performance of sponge-based moving bed biofilm reactor. *Bioresour. Technol.* **2016**, *219*, 762–767. [CrossRef] [PubMed]

12. Accinelli, C.; Saccà, M.L.; Mencarelli, M.; Vicari, A. Application of bioplastic moving bed biofilm carriers for the removal of synthetic pollutants from wastewater. *Bioresour. Technol.* **2012**, *120*, 180–186. [CrossRef] [PubMed]

13. Borges, M.E.; Sierra, M.; Cuevas, E.; García, R.D.; Esparza, P. Photocatalysis with solar energy: Sunlight-responsive photocatalyst based on TiO2 loaded on a natural material for wastewater treatment. *Sol. Energy* **2016**, *135*, 527–535. [CrossRef]

14. Wang, S.; Peng, Y. Natural zeolites as effective adsorbents in water and wastewater treatment. *Chem. Eng. J.* **2010**, *156*, 11–24. [CrossRef]

15. Cruz-Morató, C.; Lucas, D.; Llorca, M.; Rodriguez-Mozaz, S.; Gorga, M.; Petrovic, M.; Barceló, D.; Vicent, T.; Sarrà, M.; Marco-Urrea, E. Hospital wastewater treatment by fungal bioreactor: Removal efficiency for pharmaceuticals and endocrine disruptor compounds. *Sci. Total Environ.* **2014**, *493*, 365–376. [CrossRef] [PubMed]

16. Zhao, Y.; Cao, D.; Liu, L.; Jin, W. Municipal Wastewater Treatment by Moving-Bed-Biofilm Reactor with Diatomaceous Earth as Carriers. *Water Environ. Res.* **2006**, *78*, 392–396. [CrossRef] [PubMed]

17. Bhuptawat, H.; Folkard, G.; Chaudhari, S. Innovative physico-chemical treatment of wastewater incorporating Moringa oleifera seed coagulant. *J. Hazard. Mater.* **2007**, *142*, 477–482. [CrossRef] [PubMed]

18. Laidani, Y.; Hanini, S.; Henini, G. Use of fiber Luffa cylindrica for waters traitement charged in copper. Study of the possibility of its regeneration by desorption chemical. *Energy Procedia* **2011**, *6*, 381–388. [CrossRef]

19. Henini, G.; Laidani, Y.; Souahi, F.; Hanini, S. Study of static adsorption system phenol/Luffa cylindrica fiber for industrial treatment of wastewater. *Energy Procedia* **2012**, *18*, 395–403. [CrossRef]

20. Li, S.; Liu, F.; Su, Y.; Shao, N.; Yu, D.; Liu, Y.; Liu, W.; Zhang, Z. Luffa sponge-derived hierarchical meso/macroporous boron nitride fibers as superior sorbents for heavy metal sequestration. *J. Hazard. Mater.* **2019**, *378*, 120669. [CrossRef]

21. Altınışık, A.; Gür, E.; Seki, Y. A natural sorbent, Luffa cylindrica for the removal of a model basic dye. *J. Hazard Mater* **2010**, *179*, 658–664. [CrossRef] [PubMed]

22. Iqbal, M.; Edyvean, R. Alginate coated loofa sponge discs for the removal of cadmium from aqueous solutions. *Biotechnol. Lett.* **2004**, *26*, 165–169. [CrossRef]

23. Liu, Z.; Pan, Y.; Shi, K.; Wang, W.; Peng, C.; Li, W.; Sha, D.; Wang, Z.; Ji, X. Preparation of hydrophilic luffa sponges and their water absorption performance. *Carbohydr. Polym.* **2016**, *147*, 178–187. [CrossRef] [PubMed]

24. Hideno, A.; Ogbonna, J.C.; Aoyagi, H.; Tanaka, H. Acetylation of loofa (*Luffa cylindrica*) sponge as immobilization carrier for bioprocesses involving cellulase. *J. Biosci. Bioeng.* **2007**, *103*, 311–317. [CrossRef] [PubMed]

25. Do, T.T.; Tran, L.V.; Bui, T.K.V.; Nguyen, T.Q. *Research on the Application of Natural Materials as Biological Carriers in Wastewater Treatment for Beer Industry Using MBBR Technology*; Scientific Research Report; Thuy Loi University: Hanoi, Vietnam, 2016. (In Vietnamese)

26. Panneerdhass, R.; Gnanavelbabu, A.; Rajkumar, K. Mechanical properties of luffa fiber and ground nut reinforced epoxy polymer hybrid composites. *Procedia Eng.* **2014**, *97*, 2042–2051. [CrossRef]

27. Rice, E.W.; Baird, R.B.; Eaton, A.D.; Clesceri, L.S. *Standard Methods for the Examination of Water and Wastewater*; American Public Health Association: Washington, DC, USA, 2012.

28. Muyzer, G.; De Waal, E.C.; Uitterlinden, A.G. Profiling of complex microbial populations by denaturing gradient gel electrophoresis analysis of polymerase chain reaction-amplified genes coding for 16S rRNA. *Appl. Environ. Microbiol.* **1993**, *59*, 695–700. [CrossRef] [PubMed]

29. Schwieger, F.; Tebbe, C.C. A new approach to utilize PCR–single-strand-conformation polymorphism for 16S rRNA gene-based microbial community analysis. *Appl. Environ. Microbiol.* **1998**, *64*, 4870–4876. [CrossRef] [PubMed]

30. Yuan, Q.; Wang, H.; Hang, Q.; Deng, Y.; Liu, K.; Li, C.; Zheng, S. Comparison of the MBBR denitrification carriers for advanced nitrogen removal of wastewater treatment plant effluent. *Environ. Sci. Pollut. Res.* **2015**, *22*, 13970–13979. [CrossRef] [PubMed]

31. The Ministry of Natural Resources and Environment. *QVCN40:2011/BTNMT: National Technical Regulation on Industrial Wastewater (In Vietnamese)*; MONRE: Cầu Giấy, Vietnam, 2011.

32. Martín-Pascual, J.; López-López, C.; Cerdá, A.; Gonzaález-Loópez, J.; Hontoria, E.; Poyatos, J.M. Comparative Kinetic Study of Carrier Type in a Moving Bed System Applied to Organic Matter Removal in Urban Wastewater Treatment. *Water Air Soil Pollut.* **2012**, *223*, 1699–1712. [CrossRef]

33. Barwal, A.; Chaudhary, R. To study the performance of biocarriers in moving bed biofilm reactor (MBBR) technology and kinetics of biofilm for retrofitting the existing aerobic treatment systems: A review. *Rev. Environ. Sci. Bio.* **2014**, *13*, 285–299. [CrossRef]

34. Im, J.-H.; Woo, H.-J.; Choi, M.-W.; Han, K.-B.; Kim, C.-W. Simultaneous organic and nitrogen removal from municipal landfill leachate using an anaerobic-aerobic system. *Water Res.* **2001**, *35*, 2403–2410. [CrossRef]

35. Sánchez, E.; Borja, R.; Travieso, L.; Martín, A.; Colmenarejo, M.F. Effect of organic loading rate on the stability, operational parameters and performance of a secondary upflow anaerobic sludge bed reactor treating piggery waste. *Bioresour. Technol.* **2005**, *96*, 335–344. [CrossRef]

36. Yoshino, Y.; Nakazawa, S.; Otani, S.; Sekizuka, E.; Ota, Y. Nosocomial bacteremia due to Kluyvera cryocrescens: Case report and literature review. *IDCases* **2016**, *4*, 24–26. [CrossRef] [PubMed]

37. Iyer, R.; Iken, B.; Damania, A. Whole genome of Klebsiella aerogenes PX01 isolated from San Jacinto River sediment west of Baytown, Texas reveals the presence of multiple antibiotic resistance determinants and mobile genetic elements. *Genom. Data* **2017**, *14*, 7–9. [CrossRef] [PubMed]

38. Vishnivetskaya, T.A.; Kathariou, S.; Tiedje, J.M. The Exiguobacterium genus: Biodiversity and biogeography. *Extremophiles* **2009**, *13*, 541–555. [CrossRef] [PubMed]

39. Kulshreshtha, N.M.; Kumar, A.; Dhall, P.; Gupta, S.; Bisht, G.; Pasha, S.; Kumar, R. Neutralization of alkaline industrial wastewaters using *Exiguobacterium* sp. *Int. Biodeter. Biodegr.* **2010**, *64*, 191–196. [CrossRef]

40. Taguchi, M.S.; Hanai, Y. New Wastewater treatment solution using Bacillus. *Fuji Electr. Rev.* **2017**, *63*, 54–59.

41. Dastager, S.G.; Mawlankar, R.; Sonalkar, V.V.; Thorat, M.N.; Mual, P.; Verma, A.; Krishnamurthi, S.; Tang, S.-K.; Li, W.-J. Exiguobacterium enclense sp. nov., isolated from sediment. *Int. J. Syst. Evol. Micr.* **2015**, *65*, 1611–1616. [CrossRef] [PubMed]

42. Poffe, R.; de Beeck, E.O. Enumeration of Aeromonas hydrophila from domestic wastewater treatment plants and surface waters. *J. Appl. Bacteriol.* **1991**, *71*, 366–370. [CrossRef] [PubMed]

43. Azizi, S.; Valipour, A.; Sithebe, T. Evaluation of different wastewater treatment processes and development of a modified attached growth bioreactor as a decentralized approach for small communities. *Sci. World J.* **2013**, *2013*, 156870. [CrossRef] [PubMed]

44. Grimont, F.; Grimont, P.A. The genus enterobacter. In *The Prokaryotes*; Dworkin, M., Falkow, S., Eds.; Springer: New York, NY, USA, 2006; Volume 6, pp. 197–214.

45. Abbas, S.Z.; Rafatullah, M.; Ismail, N.; Lalung, J. Isolation, identification, characterization, and evaluation of cadmium removal capacity of Enterobacter species. *J. Basic Microbiol.* **2014**, *54*, 1279–1287. [CrossRef] [PubMed]

46. Varghese, R. Bioaccumulation of cadmium by *Pseudomonas* sp. isolated from metal polluted industrial region. *Environ. Res. Eng. Manag. J.* **2012**, *61*, 58–64. [CrossRef]

47. Hoang, V.; Delatolla, R.; Abujamel, T.; Mottawea, W.; Gadbois, A.; Laflamme, E.; Stintzi, A. Nitrifying moving bed biofilm reactor (MBBR) biofilm and biomass response to long term exposure to 1 C. *Water Res.* **2014**, *49*, 215–224. [CrossRef] [PubMed]

 *sustainability*

Article

The Potential Role of Hybrid Constructed Wetlands Treating University Wastewater—Experience from Northern Italy

Stevo Lavrnić [1], **Maribel Zapater Pereyra** [2], **Sandra Cristino** [3,*] , **Domenico Cupido** [4],
Giovanni Lucchese [4], **Maria Rosaria Pascale** [3], **Attilio Toscano** [1] and **Maurizio Mancini** [4]

1   Department of Agricultural and Food Sciences, Alma Mater Studiorum-University of Bologna,
    Viale Fanin 50, 40127 Bologna, Italy; stevo.lavrnic@unibo.it (S.L.); attilio.toscano@unibo.it (A.T.)
2   Independent Researcher, Gottfried-Keller-Str. 25, 81245 Munich, Germany; maribel_zapater@hotmail.com
3   Department of Biological, Geological and Environmental Sciences,
    Alma Mater Studiorum-University of Bologna, Via San Giacomo 12, 40126 Bologna, Italy;
    mariarosaria.pascal2@unibo.it
4   Department of Civil, Chemical, Environmental, and Materials Engineering,
    Alma Mater Studiorum-University of Bologna, Via Umberto Terracini 28, 40131 Bologna, Italy;
    domenicocupido1991@gmail.com (D.C.); giovannilucchesefrancesco@gmail.com (G.L.);
    maurizio.mancini@unibo.it (M.M.)
*   Correspondence: sandra.cristino@unibo.it

Received: 10 November 2020; Accepted: 14 December 2020; Published: 18 December 2020

Abstract: University wastewater is a type of wastewater with higher pollutants load and flow rate variability than typical domestic wastewater. Constructed wetlands (CW) could be used for university wastewater treatment and consequently for wastewater reuse. A hybrid CW pilot plant, at the University of Bologna (Italy), was monitored to assess its potential to be used at the university. Its treatment performance was monitored for one year and public acceptance explored through a survey. The pilot plant had two treatment lines, (1) a vertical flow CW (VFCW) and a planted horizontal flow CW (HFCW), and (2) the same VFCW and an unplanted horizontal flow filter (HFF). The HFCW achieved higher removals than the HFF, but it was also found to be prone to higher water losses. However, both treatment lines met the Italian limits for discharge in natural water bodies and some of the limits for wastewater reuse in Italy and the EU. The VFCW alone was not able to meet the same limits, demonstrating the advantages of hybrid over single stage CWs. A positive attitude towards CWs and wastewater reuse was found among the survey participants. Therefore, hybrid CWs (planted and unplanted) are considered a feasible technology for application at universities.

Keywords: hybrid constructed wetland; public acceptance; wastewater reuse; wastewater treatment

---

1. Introduction

Climate change and human activities can have negative effects on the environment [1,2], and nature-based solutions (NBS), the low-cost and green technologies, are gaining more attention in the recent years due to their numerous benefits and the ability to mitigate some of these negative effects [2,3]. Constructed wetlands (CWs), a type of NBS, are engineered systems that use the processes occurring in natural wetlands in a more controlled environment in order to treat wastewater [4,5]. They can be considered as a good solution for small or medium communities [6,7] since they can be used for different types of wastewater [8,9] and, therefore, might also be convenient for application in university campuses. Furthermore, CWs can also provide different ecosystem services such as aesthetic, educational and recreational benefits [10], convenient for the public spaces such as those of universities.

CWs can produce effluent suitable for reuse [11]; however, in order to use these systems as a part of wastewater reuse scheme on a larger scale and in countries with strict reuse limits, their performance has to be further improved. For example, it was reported that nitrogen removal in CWs might be low due to its limited capacity for denitrification and nitrification, and that removal of pathogens, one of the most important parameters for wastewater reuse, could often be unsatisfactory [2,12]. In addition, CWs can have large area requirements [8] and an increased water loss due to evapotranspiration that can have a considerable influence on their performance [13].

Single stage CWs cannot always reach strict water quality standards and therefore improvements (e.g., recirculation, artificial aeration and innovative media) or a combination of two or more CWs (hybrid systems) are commonly used, in order to enhance the removal efficiency, especially that of nitrogen [2,4]. For instance, Lavrnić and Mancini (2016) [12] showed that hybrid CWs in Southern Europe treating raw or primary treated municipal wastewater reached standards for the reuse more often than single stage systems.

Isolated buildings with green areas, such as university campuses, might need a decentralised wastewater treatment system, and, at the same time, they offer possibilities for wastewater reuse, for example, in irrigation. However, the quality of wastewater coming from the university buildings might be considerably different from domestic wastewater due to the absence of sources such as showers or clothes washing that might represent even 50% of the residential water consumption [14]. Moreover, that type of wastewater can be characterised by high inflow variability due to the low affluence of students and university staff during weekends, exam periods and holidays [15]. The presence of those factors can certainly be a challenge and, therefore, it is important to estimate the possibility for application of CWs in these conditions.

To achieve wider application of wastewater reuse practices, it is not enough to focus only on the water quality, but different aspects should be analysed together [16]. For example, public sites with CWs that were receiving primarily treated municipal wastewater were found to have a lower number of visitors than similar sites (with CWs or other natural treatment systems) receiving cleaner influent [17]. Moreover, since understanding CWs and the concept of treated wastewater reuse might not be very common among the general population, certain resistance to their application is usually encountered. In fact, public perception was reported to be one of the main problems preventing successful application of water reuse projects [18]. On the other hand, it was found that university students in Germany generally have positive attitudes towards water reuse, but at the same time they had many doubts about its quality [19].

Therefore, the objective of this research was to evaluate the potential of a hybrid CW to be used as a wastewater treatment technology at the University of Bologna (Northern Italy). For that purpose, two hybrid systems (with and without plants) were compared against each other and versus a single stage CW to understand their performance and effluent reuse capacity according to the current legislations. In addition, the attitude of the users towards CWs and wastewater reuse in their immediate surrounding was also explored.

2. Materials and Methods

2.1. Experimental Set-Up

The pilot plant used in this research (Figure 1a) was located at the School of Engineering and Architecture of the University of Bologna (Italy). It consisted of a septic tank (primary treatment), an inflow tank, a VFCW (secondary treatment) and two systems as tertiary treatment—a horizontal flow filter (HFF, not planted) and a horizontal flow CW (HFCW, planted) (Figure 1b).

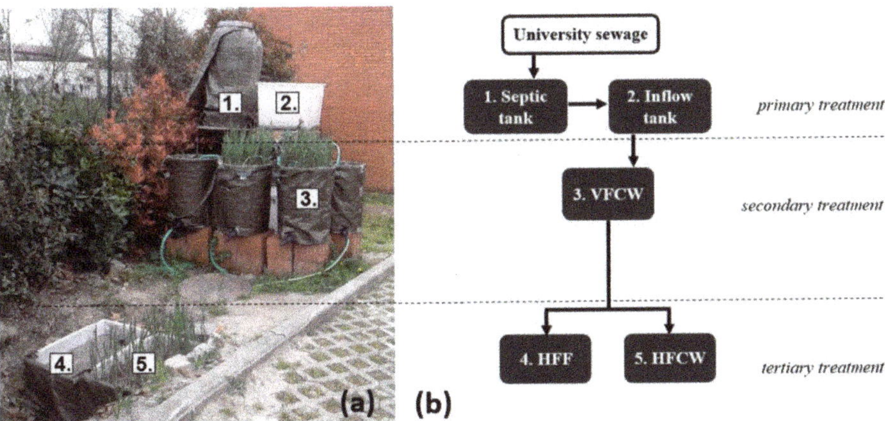

Figure 1. A photo of the pilot plant taken in March 2016 (**a**) and its schematic representation (**b**). VFCW—Vertical flow constructed wetland, HFF—Horizontal flow filter, HFCW—Horizontal flow constructed wetland.

The pilot plant was built in April 2015, and it treated wastewater coming from a university building (mostly blackwater) and an attached cafeteria. The VFCW and HFCW were planted with *Phragmites australis* at a density of 8 plants m^{-2} and filled with tap water for a period of 3 weeks to facilitate the root development. The start-up phase was 4 months long (June–October 2015), during which the systems were fed with a mixture of tap water and wastewater (in June) before being completely irrigated with only wastewater (July–September). This was a similar start-up phase as the one used by other studies [20–22].

The VFCW, which was used in a previous research to test the effect of plants and earthworms on wastewater treatment [23,24], was filled with 12 cm of gravel at the bottom (Ø 0.63–5 cm), while the top 33 cm of substrate was sand (Ø 2–4 mm). Both HFF and HFCW had the dimensions of 69 cm (length), 41.5 cm (width) and 35 cm (depth). The main substrate used was sand (Ø 1–4 mm), but the effluent pipe was covered with gravel (Ø 0.63–5 cm) in order to facilitate the drainage.

The pilot plant was monitored for a period of one year (October 2015–October 2016). During the cold period of the year (October–April), the horizontal mesocosms received 6 L twice a week and the retention time was ~4.5 days. Conversely, during the warm period of the year (April–October), they received 6 L of water 3 times a week and the retention time was ~3.75 days. In addition, during the warmest period (the second half of July and the whole of September; operation of the pilot plant was suspended in August due to the university closure) the inflow was increased to 8 L three times a week, with the retention time unchanged, in order to prevent drying due to high evapotranspiration rate. Consequently, the hydraulic loading rate varied between 6 and 12 mm d^{-1}.

2.2. Wastewater Sampling and Analytical Methods

Influents and effluents from the pilot plant were tested for different water quality parameters approximately every 3–4 weeks in the period of October 2015–October 2016 (Table 1). The analyses were performed following APHA (2005) [25] unless stated otherwise. Samples were analysed for pH by the electrometric method, chemical oxygen demand (COD) spectrophotometrically with a COD Vario cuvette kit (Aqualytic, Germany) and total suspended solids (TSS) by the gravimetric method. Total nitrogen (TN) and total phosphorus (TP) were analysed by digestion by the persulfate method followed by measurements of NO$_3^-$-N (ultraviolet spectrophotometric screening method) and PO$_4^{3-}$-P (vanadomolybdophosphoric acid colorimetric method), respectively. Finally, different ions such as nitrate (NO$_3^-$), nitrite (NO$_2^-$), phosphate (PO$_4^{3-}$), chloride (Cl$^-$), bromide (Br$^-$), and sulphate (SO$_4^{2-}$) were analysed by ion chromatography (DX-120, Dionex Corporation, CA, USA).

Table 1. Performance of the pilot plant during the period October 2015–October 2016 (mean ± st. error (sample size)).

Parameter	Influent [a]	VFCW Effluent [b]		HFF Effluent		HFCW Effluent		*t* Test [e]
-	Value	Value	RE [c] (%)	Value	RE [d] (%)	Value	RE [d] (%)	*p* Value
pH	6.65 (7)	7.18 (12)	-	7.27 (10)	-	6.95 (9)	-	-
COD (mg L^{-1})	886 ± 74 (13)	263 ± 32 (15)	70.4	151 ± 29 (15)	42.5	157 ± 30 (14)	40.1	0.953
TSS (mg L^{-1})	168 ± 28 (10)	33 ± 4 (18)	80.4	9 ± 1 (18)	73.2	9 ± 2 (16)	72.7	0.412
TN (mg L^{-1})	65 ± 6 (18)	33 ± 6 (17)	49.3	13 ± 3 (17)	59.8	4 ± 1 (10)	88.8	0.003
NO_3^--N (mg L^{-1})	0.85 ± 0.12 (19)	1.16 ± 0.60 (18)	-	5.20 ± 1.35 (18)	-	1.16 ± 0.38 (16)	-	0.009
NO_2^--N (mg L^{-1})	0.03 ± 0.03 (19)	0.04 ± 0.04 (18)	-	2.86 ± 2.79 (18)	-	3.39 ± 3.39 (16)	-	0.146
TP (mg L^{-1})	12.32 ± 2.83 (17)	6.50 ± 1.18 (15)	47.3	2.48 ± 0.70 (17)	61.8	0.75 ± 0.19 (14)	88.5	0.008
PO_4^{3-}-P (mg L^{-1})	7.63 ± 1.43 (17)	4.93 ± 0.83 (15)	34.2	2.03 ± 0.13 (18)	58.9	0.24 ± 0.11 (13)	95.1	<0.001
Cl^- (mg L^{-1})	78 ± 5 (19)	77 ± 7 (18)	0	74 ± 7 (18)	5	70 ± 15 (15)	9.7	0.396
Br^- (mg L^{-1})	2.56 ± 0.35 (19)	1.72 ± 0.26 (18)	33	1.87 ± 0.24 (18)	-	1.93 ± 0.24 (16)	-	0.373
SO_4^{2-} (mg L^{-1})	69 ± 5 (19)	66 ± 9 (18)	3.5	75 ± 8 (18)	-	59 ± 19 (15)	10.2	0.031
E. coli (log$_{10}$ CFU 100 mL^{-1})	5.69 ± 5.02 (10)	5.09 ± 4.56 (10)	74.7	3.25 ± 3.03 (14)	98.5	2.54 ± 2.23 (13)	99.7	0.133
Total coliforms (log$_{10}$ CFU 100 mL^{-1})	5.85 ± 5.81 (4)	4.81 ± 4.40 (10)	90.7	3.84 ± 3.63 (11)	89.3	3.63 ± 3.46 (10)	93.5	0.081
Enterococcus (log$_{10}$ CFU 100 mL^{-1})	5.51 ± 4.59 (10)	5.21 ± 4.76 (10)	50.1	2.92 ± 2.49 (14)	99.5	2.06 ± 1.90 (13)	99.9	0.015

[a] Sedimentation tank effluent; [b] The VFCW effluent was the influent for both HFF and HFCW; [c] Removal efficiency (RE) was calculated based on the influent (sedimentation tank effluent); [d] RE was calculated based on the VFCW effluent; [e] T test *p*-values show the statistical comparison of the effluents HFF and HFCW. The conducted tests were *t* test or Mann–Whitney U test. Bolded values show significant differences ($p < 0.05$). VFCW—Vertical flow constructed wetland, HFF—Horizontal flow filter, HFCW—Horizontal flow constructed wetland.

The microbiological parameters (*E. coli*, Total coliforms and *Enterococcus*) were analysed by the membrane filter technique followed by incubation and enumeration using a Chromogenic Coliform Agar for *E. coli* and Total coliforms [26], and a Slanetz Bartley Agar for *Enterococcus* [27].

2.3. Effect of Plants on Wastewater Treatment

The effluent concentrations of the HFCW and the HFF were statistically compared (*t* test) in order to evaluate the contribution of plants to wastewater treatment during the whole experimental period. The data were first checked for normality and equal variance, and if the assumptions were not met, the values were log_{10} transformed. In the case when the assumptions could not be met even after the transformation, a Mann–Whitney U test was used. All the analyses were performed using SigmaPlot 13 software.

2.4. Public Acceptance of CWs and Wastewater Reuse

An online questionnaire was created to find out the attitude of the daily users of the building of the School of Engineering and Architecture towards CWs and a possible reuse of wastewater treated. The participants were asked seven questions about their opinion of the mentioned topic, apart from questions regarding their social structure (sex, age, level of education). Answers were given by using the scale 1–5, 1 being the most negative and 5 the most positive attitude. The contestants were mostly reached through the social network Facebook.

3. Results and Discussion

3.1. Treatment and Reuse Potential

Results of the pilot plant performance are given in Table 1. The primarily treated university wastewater (septic tank effluent) will be considered as "influent" for the purpose of this study. It can be seen that, even after the primary treatment, wastewater can be considered to be of high strength [28]. That was due to the fact that wastewater was coming mostly from university toilets and cafeteria, so the factors that usually dilute domestic wastewater (e.g., showers, washing machines) were not present.

The majority of pollutants were mostly removed in the VFCW, as discussed in more detail in Lavrnić et al. (2019) [24]. Similarly, Ávila et al. (2016) [29] and Zhai et al. (2016) [30] also found that majority of pollutants were mostly removed in the first stage of a hybrid CW (VFCW+HFCW). Such behaviour might be explained by first-order kinetics and the increased removal in the case of high influent concentrations [31].

The pollutants removal achieved by both lines (HFCW and HFF) of the pilot plant was high. For example, in both treatment lines, the COD decreased from 886 to <158 mg L^{-1} and TSS from 168 to 9 mg L^{-1} (Table 1). Nutrients removal efficiencies were higher than 79% for both TN and TP. In addition, all three indicator pathogens were removed for more than 98%, with *E. coli* decreasing for more than 2.3 and *Enterococcus* for more than 2.5 log units (Table 1). These results are comparable or higher than those reported in Torrijos et al. (2016) [32] and Zhai et al. (2016) [30], although influents used in those studies (e.g., COD in the range 193–405 mg L^{-1}, TN in the range 53–57 mg L^{-1}) were of much lower strength than the one used in this manuscript. Therefore, the high removal achieved in this study can be attributed to a longer hydraulic retention time (HRT) (>4.5 in this study vs. <2.7 days in Zhai et al. (2016)) or explained by the first-order kinetics where a high influent concentration results in a high removal efficiency [31]. Both systems were new (in operation for less than 2 years) as the hybrid CW reported here, and therefore no difference in the removal efficiencies can be attributed to the age of the system.

It should be noted that the removal efficiencies obtained during the one-year experimental period might change as the system ages, since it was suggested that some removal processes in CWs can vary with time [33]. For instance, adsorption processes were found to be responsible for TN removal at the beginning of the CW operation (approximately 5–24 months), while microbial processes gained advantage as operation time passed [34]. TP removal efficiency can also be reduced as the substrate's sorption capacity decreases over time [35]. In addition, certain processes are influenced by the presence of organic matter and since it tends to accumulate over time in CWs [36], it might cause a change in effluent concentrations over a longer experimental period.

To understand how the systems performed according to 3 legal thresholds—guidelines for discharge to natural water bodies (in Italy) and for wastewater reuse in Italy and European Union (EU). The standards for discharge to natural water bodies were not met by the VFCW effluent (COD and *E. coli* values were too high) but they were met by the HFF and HFCW effluents (Table 2) [37]. Therefore, the studied hybrid pilot plant could be used to treat university wastewater and the effluent could be discharged into surface water bodies.

Wastewater reuse in Italy does not distinguish different reuse types, but some exceptions are allowed when the treatment is done by a NBS (called "general reuse" in this study) and when the treated wastewater is intended to be reused in irrigation (called "irrigational reuse" in this study) [38]. The VFCW, HFF and HFCW produced effluents that had the values of COD and *E. coli* too high to meet both general and irrigational reuse (Table 2). In addition, the HFF also had a TP concentration (2.48 mg L^{-1}) slightly above the threshold for general reuse (2 mg L^{-1}) but met the limits for irrigational reuse (10 mg L^{-1}) (Table 2). Hence, the effluent of both systems should be further treated if planned to be reused in Italy.

The EU regulations for water reuse in agricultural irrigation [39] were recently adopted by the EU (approved in May 2020, valid from June 2023). They are expected to represent a boost for this practice in the EU, and in particular in Italy, since they are more flexible than the current Italian guidelines. When the effluent concentrations are compared to the new EU limits, it can be seen that TSS removal was on the satisfactory level and its concentration was below the limits for all four reuse types (Table 2). *E. coli* removal by the hybrid system was not sufficient for reuse types A (e.g., root crops consumed raw and food crops where the edible portion is in direct contact with reclaimed water) and B (e.g., food crops consumed raw where the edible portion is not in direct contact with reclaimed water, all irrigation methods). However, both HFF and HFCW effluents are suitable for Type D (industrial, energy, and seeded crops), while HFCW effluent also satisfies limits for Type C (the same as type B, but allowing only drip irrigation) [39]. It is important to mention also that the VFCW alone would not be able to reach limits for any of the four reuse types due to the high *E. coli* concentration, emphasising the benefits of hybrid systems in comparison with single stage CWs.

In order to meet the Italian reuse limits or the reuse types A and B of the new EU guidelines (Table 2), an additional treatment step would be required. Ávila et al. (2013) [40] used a free water surface CW (FWSCW) as the additional treatment step after a hybrid CW (VFCW+HFCW), and reported that it did not considerably improve the system's performance. Wu et al. (2016) [41] concluded that, in general, HFCWs have a higher removal rates of pathogens than FWSCWs and therefore, adding another HFF or HFCW at the end of the studied treatment plant in Bologna would probably help to meet the reuse limits. Other solutions include a combination of the treatment line with an ultraviolet disinfection unit disinfection unit [42] or introduction of aeration in the horizontal system [41], both of which can increase removal of pathogens and probably help meeting the reuse limits [2]. Moreover, a combination of CWs with more intensive technologies such as up-flow anaerobic sludge blanket were also successfully used for wastewater treatment of smaller communities [43].

Regarding effluent water availability, the studied HFCW sometimes turned to a zero-discharge system during the summer months, probably due to the presence of plants and their transpiration processes. The HFF, on the other hand, was never found to be completely dry. Milani and Toscano (2013) [13] concluded that evapotranspiration rates for small-scale CWs can be much higher compared to natural wetlands with larger areas due to the clothesline and oasis effect. High water loss through evapotranspiration was also reported by Zapater-Pereyra et al. (2016) [44] for a rooftop wetland with a small depth (9 cm) and a large area (306 m^2). Unfortunately, in the current study it was not possible to estimate the exact evapotranspiration losses. However, pronounced water loses probably occurred due to the small flow rate and scale of the pilot plant. Nevertheless, further tests on a bigger scale would be needed to test this hypothesis.

Table 2. Comparison of the studied systems effluent with the standards for reuse and discharge to natural water bodies.

Parameter	VFCW	HFF	HFCW	Italy Discharge Regulation [37]	Italy Reuse Regulation [a] [38]		EU Regulation for Water Reuse [b] [39]			
-	Mean Value	Mean Value	Mean Value	Natural Water Bodies	General	Irrigational	Type A	Type B	Type C	Type D
COD (mg L^{-1})	263	151	157	160	100	100	-	-	-	-
TSS (mg L^{-1})	33	9	9	80	10	10	10	35	35	35
TN (mg L^{-1})	33	13	4	-	15	35	-	-	-	-
TP (mg L^{-1})	7	2.48	0.75	10	2	10	-	-	-	-
E. coli (CFU 100 mL^{-1})	123,400	1790	350	5000 [c]	50	100	10	100	1000	10,000

[a] The Italian guideline for reuse does not include different types of reuse, but it does allow some variations in their reuse limits when the technology used is a nature-based system (called "general reuse" in this study), and also when treated wastewater is intended to be used in irrigation (called "irrigational reuse" in this study); [b] Type A—All food crops, including root crops consumed raw and food crops where the edible portion is in direct contact with reclaimed water; Type B, C—Food crops consumed raw where the edible portion is produced above ground and is not in direct contact with reclaimed water, Processed food crops, Non-food crops including crops to feed milk- or meat-producing animals, depending on the irrigation method (type B allows all irrigation methods, type C only drip irrigation); Type D—Industrial, energy, and seeded crops. [c] Maximal recommended value, the competent authority can decide the exact limit.

3.2. Hybrid CWs for Wastewater Reuse

As previously stated, single stage CWs are not as effective as hybrid ones, and therefore, they normally have more difficulties to meet reuse limits [2,12]. For example, Herrera Melián et al. (2010) [45] concluded that a second stage of a hybrid CW (VFCW+HFCW) performed at least equally as the first one regarding disinfection performance and that together they represented a robust option for wastewater treatment and reuse. Similarly, Avila et al. (2013) [40]concluded that a second and even a third stage of a hybrid CW would probably prove crucial to achieve an effluent quality needed for reuse. Moreover, Zurita and White (2014) [11] treated university wastewater, with a similar strength to the one used in this study, and showed that, although the first stage of a hybrid CW had high removal efficiencies, at least two stages of treatment are needed to achieve the disinfection levels required for the reuse of treated wastewater for irrigation.

Similar results were obtained in the current study when treating university wastewater. Despite the fact that the VFCW did most of the treatment, it was not enough to reach either reuse or discharge limits (Table 1). Hence, for such a purpose, a hybrid system is more suitable and recommended.

However, legislations should be considered as the real bottleneck, since, in the case of flexible standards (i.e., the new EU regulation) any of the hybrid systems (VFCW+HFCW or VFCW+HFF) were able to meet the reuse standards, while in the case of strict reuse limits (i.e., Italian regulation), it is recommended to have an additional step (after the hybrid system) mainly to meet the pathogen limitations.

3.3. Effect of Plants on Wastewater Treatment

To understand the effect of plants, a planted CW (HFCW) and an unplanted filter (HFF) were tested as the second stage of the hybrid system for a period of 1 year. It should be noted that, although reaching steady state conditions in a CW might require between 1 and even 4 years [33,46], several CW studies were conducted during an experimental period similar to the one used in this study [46–48].

Along the whole experimental period, the HFCW showed to have significantly lower concentrations ($p < 0.05$) for several parameters (e.g., TN, TP, and *Enterococcus*) than the HFF (Table 1). Carballeira et al. (2016) [47] also reported that plants had a positive effect on the nutrient removal in horizontal systems that were monitored for 2.5 years. However, no significant difference was found ($p > 0.05$) in COD or TSS effluent concentrations between the HFCW and HFF in the present study (Table 1). That is in accordance with Ciria et al. (2005) [21] and Collison and Grismer (2015) [46], who did not find any difference in COD and TSS removal between planted and unplanted systems over an experimental period of 18 and 7 months, respectively. Although most of the oxygen introduced to the substrate by plants is used for respiration, usually some of it is lost to the rhizosphere, and hence is not sufficient for aerobic degradation [3]. It can be hypothesised that the amount of oxygen introduced to the HFCW substrate by the plants was not enough to foster aerobic degradation of organic matter that would increase COD and TSS removal beyond the level reached in the HFF, causing anaerobic pathways to be dominant for organic matter removal [49].

The observed differences in nutrient removal between HFF and HFCW (effluent concentration of TN: 13 vs. 4 mg L^{-1} and TP: 2.48 vs. 0.75 mg L^{-1}, respectively) (Table 1) can be attributed to the presence of plants. They were reported to have an important role in the removal of nutrients [9,47] due to the direct nutrient absorption, the action of the aerobic microorganisms located in their rhizosphere [50] or the provision of an additional source of carbon for denitrification processes due to plant decay [51].

Among the Cl$^-$, Br$^-$ and SO$_4$$^{2-}$ ions, the presence of plants significantly increased the SO$_4$$^{2-}$ removal showing statistically significant differences ($p < 0.05$) (Table 1) between the HFCW and the HFF. That is probably connected to the increased presence of different carbon sources from the plant litter that enabled SO$_4$$^{2-}$ reduction [52].

It is not clear if vegetation contributes to the removal of indicator bacteria [53]. For example, Wu et al. (2016) [41] stated that in the majority of cases, the presence of vegetation has a positive influence on the removal of indicator bacteria in HFCWs. Headley et al. (2013) [48] suggested, based on a 16-month long experimental period, that plants do not play a big role in that process since removal of pathogens mostly depends on physical parameters. Carballeira et al. (2016) [47] showed that there is often little or no significant difference in pathogens removal between planted and unplanted systems. In this study, the HFCW had a significantly lower effluent concentration of *Enterococcus* compared to the HFF (2.06 vs. 2.92 $\log_{10}$ CFU 100 mL^{-1}, respectively; Table 1), but the authors could not find any similar situation in the published literature. However, it has been indicated that the presence of plants can, although with a minor effect, improve pathogen removal [47] maybe due to the physical effects (changing the system hydraulics) or through biological ones (increased surface area availability at plant roots or root exudates) [41]. Therefore, it is hypothesised that *E. coli* and total coliforms removal was not affected by the presence of plants since it was more related to the physical properties (such as HRT that was the same in both hybrid systems), while the effect of plants was more expressed, and therefore visible, in the removal of *Enterococcus*. Further research is needed in this regard.

The results about the role of plants obtained by this study are based on a one-year long experimental period. Literature provides conflicting information about the effect that different CW's operation times can have on the role of the vegetation and on the treatment performance, and therefore, these results should be taken with certain caution. For instance, while some authors reported that the removal of organic matter in CWs can be taken stable over the years [54,55] and that TN and TP removal after 10 years of operation were comparable to those recorded at the beginning [54], it was also suggested that CWs can experience important removal efficiency variations over their life time [33]. As they are aging, the HFF and HFCW might experience certain process shifts and removal rates changes, and it is not clear if and how the role of plants might vary during this time.

3.4. Public Acceptance of CWs and Their Effluent Reuse

The online questionnaire reached 76 participants and majority of them were male (59% vs. 41% of female) and students (91% vs. 9% of different types of university employees). Moreover, out of the four age groups (<29, 30–49, 50–64, >65 years old), the majority of participants (95%) was under 29 years of age. The results are given in Table 3.

Table 3. Online questionnaire and summary of the results obtained from 76 users.

Do you know:	yes	no			
• what constructed wetlands are?	61%	39%			
• that at the university there is a constructed wetland pilot plant?	34%	66%			
On a scale from 1 to 5 *, how much do you agree with:	**1**	**2**	**3**	**4**	**5**
• reusing treated wastewater for irrigation?	1%	5%	7%	28%	59%
• treating university wastewater with the existing pilot plant system?	1%	1%	14%	20%	63%
• using the water from the constructed wetland system to irrigate the university green areas?	1%	5%	8%	16%	70%
• using treated wastewater for toilets flushing in the university building?	1%	3%	8%	12%	76%

* The scale range is from 1 (do not agree at all) to 5 (very much agree).

Although 61% of the participants knew what CWs are, only 34% of them knew about the existence of the pilot plant at the School of Engineering and Architecture, University of Bologna (Italy). Moreover, the great majority of the interviewed users was in favour of treating wastewater with the pilot plant and reuse of its effluent for irrigation of green areas or toilet flushing. These results were unexpected as public approval for wastewater reuse is limited due to emotional discomfort [56]. Although the School of Engineering and Architecture offers different engineering studies that are not all connected to the environmental area, the high support might, due to similar interests, be a result of the big number of participants being from the environmental field. Also, the high educational level of people that filled the questionnaire could explain this behaviour, since it was reported to be connected with higher approval of wastewater reuse [18]. The additional factor might be the fact that the participants had somewhat direct contact with the CW systems.

The pilot plant was placed in the yard of the School of Engineering and Architecture, near the cafeteria and a green area where students would spend their breaks. Some authors of this manuscript were approached by students or staff while operating the pilot plant and were asked about the nature of their study and performance of the system. That clearly shows interest and suggests a positive attitude of the potential users towards the pilot plant, as also shown by the questionnaire results (Table 3). Ghermandi and Fichtman (2015) [17] found that a clear human involvement (e.g., through the installation of recreational facilities and educational displays) can increase interest of the general public. It suggests that a positive attitude of the participants could be maintained and increased by, for example, setting panels that would offer basic information about the system. The online questionnaire provided in this study can be considered a first step to involve the citizens, testing the compliance with wastewater treatment and reuse approach trough their feedback.

Although managers and the public are not always aware of the potential values of CWs [57], Ezeah et al. (2015) [58] reported a recent increase in public support for CWs as wastewater treatment systems due to the different benefits that they can provide. That can also explain the positive attitude of the questionnaire participants (Table 3). On the other hand, Everard et al. (2012) [59] did a survey with stakeholders associated with 16 integrated CWs (CWs that served not only for wastewater treatment, but also provided different ecosystems services) in Anne Valley catchment in South-Eastern Ireland. Although many interviewees focused on the economic benefits connected to their ability to treat wastewater, the majority of them recognised integrated CWs as socially beneficial resources that were contributing in different ways to the quality of life of the area. Moreover, it was recorded that some farmers that in the beginning had a negative opinion of CWs, accepted the idea and installed their own systems after seeing successful application at neighbouring farms [59]. That is in accordance with Rice et al. (2016) [18] who stated that people can have the initial negative opinion of the technologies they are not familiar with. Therefore, a negative or medium attitude towards the pilot plant in this questionnaire could be changed with time and different actions that would show the importance of the concept (e.g., information panels, guided visits).

4. Conclusions

The present research had the objective to assess the potential of hybrid CWs to be used in universities for wastewater treatment. The system included a septic tank, a VFCW as a secondary stage, and a HFCW (planted) and HFF (unplanted) as a tertiary stage. Both treatment lines were tested in parallel in order to determine which one is better according to the applicable reuse regulations.

The results showed that both treatment lines (VFCW+HFF and VFCW+HFCW) were able to treat wastewater to be discharged in natural water bodies. In addition, both treatment lines met the Italian reuse (general and irrigational) thresholds for TSS and TN but that was not the case for COD and *E. coli*. However, if the EU regulations are taken into consideration, both treatment lines are suitable for the reuse Type D (for industrial, energy and seeded crops) and, for instance, the effluent could be used to irrigate energy crops if universities use them for heat generation during winter. The hybrid VFCW+HFCW also satisfied the limits for the reuse Type C (e.g., for food crops consumed raw where the edible part is not in direct contact with reclaimed water) and it could even be used for irrigation of certain food crops (only with drip irrigation method) in experimental university gardens for instance.

It is worth noting that the hybrid systems had a higher pollutants removal than the VFCW alone, the later not being able to satisfy any of the reuse types mentioned or the limits for discharge to natural water bodies. On the other hand, the hybrid systems low pathogen removal can be further increased by the addition of other components such as another CW or UV lamp.

The presence of plants positively influenced the HFCW performance over the one-year study, especially for nutrient (TN and TP) and *Enterococcus* removal, compared to the HFF. For example, TN removal efficiency was 60% for the HFF and 89% for the HFCW. COD and TSS removals were independent of the presence of plants, and for both systems it was in the range 40–43% and 72–74%, respectively. However, a longer experimental period might be required to confirm these findings.

Public attitude for wastewater treatment by CWs and reuse of their effluent, an important aspect that is not always considered, was found to be very positive among the participants, a result that could indicate an increased awareness for environmental issues among young and highly educated people.

Therefore, the hybrid CWs tested (both planted and unplanted) can be considered as a viable and environmentally friendly alternative for the treatment and reuse of wastewater coming from smaller communities or individual buildings like universities.

Author Contributions: Conceptualization, S.L., M.Z.P., S.C., M.M. and A.T.; Methodology, S.L., M.Z.P., S.C. and D.C.; Formal analysis, S.L.; Investigation, S.L., D.C., G.L., S.C., M.Z.P. and M.R.P.; Resources, M.M. and S.C.; Data curation, S.L.; Writing—original draft preparation, S.L., M.Z.P., S.C., D.C. and M.M.; Writing—review and editing, S.L., M.Z.P., S.C., D.C., G.L., M.R.P., M.M. and A.T.; Supervision, M.M. and S.C. All authors have read and agreed to the published version of the manuscript.

Funding: This research received no external funding.

Conflicts of Interest: The authors declare no conflict of interest.

References

1. Pavanelli, D.; Cavazza, C.; Lavrnić, S.; Toscano, A. The Long-Term Effects of Land Use and Climate Changes on the Hydro-Morphology of the Reno River Catchment (Northern Italy). *Water* **2019**, *11*, 1831. [CrossRef]

2. Nan, X.; Lavrnić, S.; Toscano, A. Potential of Constructed Wetland Treatment Systems for Agricultural Wastewater Reuse under the EU Framework. *J. Environ. Manag.* **2020**, *275*, 111219. [CrossRef] [PubMed]

3. Pap, S.; Šolević Knudsen, T.; Radonić, J.; Maletić, S.; Igić, S.M.; Turk Sekulić, M. Utilization of Fruit Processing Industry Waste as Green Activated Carbon for the Treatment of Heavy Metals and Chlorophenols Contaminated Water. *J. Clean. Prod.* **2017**, *162*, 958–972. [CrossRef]

4. Vymazal, J. Horizontal Sub-Surface Flow and Hybrid Constructed Wetlands Systems for Wastewater Treatment. *Ecol. Eng.* **2005**, *25*, 478–490. [CrossRef]

5. Lavrnić, S.; Alagna, V.; Iovino, M.; Anconelli, S.; Solimando, D.; Toscano, A. Hydrological and Hydraulic Behaviour of a Surface Flow Constructed Wetland Treating Agricultural Drainage Water in Northern Italy. *Sci. Total Environ.* **2020**, *702*, 134795. [CrossRef] [PubMed]

6. Barbagallo, S.; Cirelli, G.L.; Marzo, A.; Milani, M.; Toscano, A. Hydraulic Behaviour and Removal Efficiencies of Two H-SSF Constructed Wetlands for Wastewater Reuse with Different Operational Life. *Water Sci. Technol. J. Int. Assoc. Water Pollut. Res.* **2011**, *64*, 1032–1039. [CrossRef] [PubMed]

7. Jucherski, A.; Nastawny, M.; Walczowski, A.; Gajewska, M.; Józ, K. Assessment of the Technological Reliability of a Hybrid Constructed Wetland for Wastewater Treatment in a Mountain Eco-Tourist Farm in Poland. *Water Sci. Technol.* **2017**, *75*, 2649–2658. [CrossRef] [PubMed]

8. Lavrnić, S.; Nan, X.; Blasioli, S.; Braschi, I.; Anconelli, S.; Toscano, A. Performance of a Full Scale Constructed Wetland as Ecological Practice for Agricultural Drainage Water Treatment in Northern Italy. *Ecol. Eng.* **2020**, *154*, 105927. [CrossRef]

9. Gorgoglione, A.; Torretta, V. Sustainable Management and Successful Application of Constructed Wetlands: A Critical Review. *Sustainability* **2018**, *10*, 3910. [CrossRef]

10. Semeraro, T.; Giannuzzi, C.; Beccarisi, L.; Aretano, R.; De Marco, A.; Pasimeni, M.R.; Zurlini, G.; Petrosillo, I. A Constructed Treatment Wetland as an Opportunity to Enhance Biodiversity and Ecosystem Services. *Ecol. Eng.* **2015**, *82*, 517–526. [CrossRef]

11. Zurita, F.; White, J.R. Comparative Study of Three Two-Stage Hybrid Ecological Wastewater Treatment Systems for Producing High Nutrient, Reclaimed Water for Irrigation Reuse in Developing Countries. *Water* **2014**, *6*, 213. [CrossRef]

12. Lavrnić, S.; Mancini, M.L. Can Constructed Wetlands Treat Wastewater for Reuse in Agriculture? Review of Guidelines and Examples in South Europe. *Water Sci. Technol. J. Int. Assoc. Water Pollut. Res.* **2016**, *73*, 2616–2626. [CrossRef] [PubMed]

13. Milani, M.; Toscano, A. Evapotranspiration from Pilot-Scale Constructed Wetlands Planted with Phragmites Australis in a Mediterranean Environment. *J. Environ. Sci. Heal. Part A Toxic Hazardous Subst. Environ. Eng.* **2013**, *48*, 568–580. [CrossRef] [PubMed]

14. Jordán-Cuebas, F.; Krogmann, U.; Asce, M.; Andrews, C.J.; Senick, J.A.; Hewitt, E.L.; Wener, R.E.; Sorensen Allacci, M.; Plotnik, D. Understanding Apartment End-Use Water Consumption in Two Green Residential Multistory Buildings. *J. Water Resour. Plan. Manag.* **2018**, *144*. [CrossRef]

15. Guedes-Alonso, R.; Montesdeoca-Esponda, S.; Herrera-Melián, J.A.; Rodríguez-Rodríguez, R.; Ojeda-González, Z.; Landívar-Andrade, V.; Sosa-Ferrera, Z.; Santana-Rodríguez, J.J. Pharmaceutical and Personal Care Product Residues in a Macrophyte Pond-Constructed Wetland Treating Wastewater from a University Campus: Presence, Removal and Ecological Risk Assessment. *Sci. Total Environ.* **2020**, *703*, 135596. [CrossRef]

16. Ricart, S. Risk-Yuck Factor Nexus in Reclaimed Wastewater for Irrigation: Comparing Farmers' Attitudes and Public Perception. *Water* **2019**, *11*, 187. [CrossRef]

17. Ghermandi, A.; Fichtman, E. Cultural Ecosystem Services of Multifunctional Constructed Treatment Wetlands and Waste Stabilization Ponds: Time to Enter the Mainstream? *Ecol. Eng.* **2015**, *84*, 615–623. [CrossRef]

18. Rice, J.; Wutich, A.; White, D.D.; Westerhoff, P. Comparing Actual de Facto Wastewater Reuse and Its Public Acceptability: A Three City Case Study. *Sustain. Cities Soc.* **2016**, *27*, 467–474. [CrossRef]

19. Schmid, S.; Bogner, F. What Germany's University Beginners Think about Water Reuse. *Water* **2018**, *10*, 731. [CrossRef]

20. Amaral, R.; Ferreira, F.; Galvão, A.; Matos, J.S. Constructed Wetlands for Combined Sewer over Fl Ow Treatment in a Mediterranean Country, Portugal. *Water Sci. Technol.* **2013**, *67*, 2739–2745. [CrossRef]

21. Ciria, M.P.; Solano, M.L.; Soriano, P. Role of Macrophyte Typha Latifolia in a Constructed Wetland for Wastewater Treatment and Assessment of Its Potential as a Biomass Fuel. *Biosyst. Eng.* **2005**, *92*, 535–544. [CrossRef]

22. Villalobos, R.M.; Zúñiga, J.; Salgado, E.; Cristina, M. Constructed Wetlands for Domestic Wastewater Treatment in a Mediterranean Climate Region in Chile. *Electron. J. Biotechnol.* **2013**, *16*, 1–13. [CrossRef]

23. Mancini, B. *Application of Sequence Based Typing (SBT) Technique to Typing Strains of Legionella spp.: Development of an Environmental Risk Map*; University of Bologna: Bologna, Italy, 2017.

24. Lavrnić, S.; Cristino, S.; Zapater-Pereyra, M.; Vymazal, J.; Cupido, D.; Lucchese, G.; Mancini, B.; Mancini, M.L. Effect of Earthworms and Plants on the Efficiency of Vertical Flow Systems Treating University Wastewater. *Environ. Sci. Pollut. Res.* **2019**, *26*, 10354–10362. [CrossRef] [PubMed]

25. APHA. *Standard Methods for the Examination of Water and Wastewater*, 21st ed.; American Public Health Association: Washington, DC, USA, 2005.

26. ISO. *9308-1 Water Quality—Enumeration of Escherichia Coli and Coliform Bacteria—Part 1: Membrane Filtration Method for Waters with Low Bacterial Background Flora*; ISO: Geneva, Switzerland, 2014.

27. ISO. *7899-2 Water Quality—Detection and Enumeration of Intestinal Enterococci—Part 2: Membrane Filtration Method*; ISO: Geneva, Switzerland, 2000.

28. Tchobanoglous, G.; Burton, E.C.; Stensel, H.D. *Wastewater Engineering: Treatment Disposal and Reuse*; Metcalf & Eddy: Bedford, MA, USA; McGraw-Hill, Inc.: Singapore, 2004.

29. Ávila, C.; García, J.; Garfí, M. Influence of Hydraulic Loading Rate, Simulated Storm Events and Seasonality on the Treatment Performance of an Experimental Three-Stage Hybrid Constructed Wetland System. *Ecol. Eng.* **2016**, *87*, 324–332. [CrossRef]

30. Zhai, J.; Xiao, J.; Rahaman, M.H.; John, Y.; Xiao, J. Seasonal Variation of Nutrient Removal in a Full-Scale Artificial Aerated Hybrid Constructed Wetland. *Water* **2016**, *8*, 551. [CrossRef]

31. Kadlec, R.H.; Wallace, S. *Treatment Wetlands*, 2nd ed.; CRC Press: Boca Raton, FL, USA, 2009.

32. Torrijos, V.; Gonzalo, O.G.; Trueba-Santiso, A.; Ruiz, I.; Soto, M. Effect of By-Pass and Effluent Recirculation on Nitrogen Removal in Hybrid Constructed Wetlands for Domestic and Industrial Wastewater Treatment. *Water Res.* **2016**, *103*, 92–100. [CrossRef]

33. Gajewska, M.; Skrzypiec, K.; Jó, K.; Mucha, Z.; Karczmarczyk, A.; Bugajski, P. Kinetics of Pollutants Removal in Vertical and Horizontal Flow Constructed Wetlands in Temperate Climate. *Sci. Total Environ.* **2020**, *718*, 137371. [CrossRef]

34. Wojciechowska, E.; Gajewska, M.; Ostojski, A. Reliability of Nitrogen Removal Processes in Multistage Treatment Wetlands Receiving High-Strength Wastewater. *Ecol. Eng.* **2017**, *98*, 365–371. [CrossRef]

35. Jóźwiakowski, K.; Bugajski, P.; Kurek, K.; De Fátima, M.; De Carvalho, N.; Adelaide, M.; Almeida, A.; Siwiec, T.; Borowski, G.; Czeka, W.; et al. The Efficiency and Technological Reliability of Biogenic Compounds Removal during Long-Term Operation of a One-Stage Subsurface Horizontal Flow Constructed Wetland. *Sep. Purif. Technol.* **2018**, *202*, 216–226. [CrossRef]

36. Reddy, G.B.; Raczkowski, C.W.; Cyrus, J.S.; Szogi, A. Carbon Sequestration in a Surface Fl Ow Constructed Wetland after 12 Years of Swine Wastewater Treatment. *Water Sci. Technol.* **2016**, *73*, 2501–2508. [CrossRef]

37. Legislative Decree, n. 152, 03/04/2006, Gazzetta Ufficiale n. 88. Environmental Regulations. Available online: https://www.gazzettaufficiale.it/dettaglio/codici/materiaAmbientale (accessed on 12 August 2020).

38. Legislative Decree, n. 185, 23/07/2003 Gazzetta Ufficiale n. 169. Italian Technical Guidelines for Wastewater Reuse. Available online: http://www.gazzettaufficiale.it/eli/id/2003/07/23/003G0210/sg (accessed on 12 August 2020).

39. European Union. *Regulation (EU) 2020/741 of the European Parliament and of the Council of 25 May 2020 on Minimum Requirements for Water Reuse (Text with EEA Relevance)*; European Union: Brussels, Belgium, 2020.

40. Ávila, C.; Garfí, M.; García, J. Three-Stage Hybrid Constructed Wetland System for Wastewater Treatment and Reuse in Warm Climate Regions. *Ecol. Eng.* **2013**, *61*, 43–49. [CrossRef]

41. Wu, S.; Carvalho, P.N.; Müller, J.A.; Manoj, V.R.; Dong, R. Sanitation in Constructed Wetlands: A Review on the Removal of Human Pathogens and Fecal Indicators. *Sci. Total Environ.* **2016**, *541*, 8–22. [CrossRef]

42. Toscano, A.; Hellio, C.; Marzo, A.; Milani, M.; Lebret, K.; Cirelli, G.L.; Langergraber, G. Removal Efficiency of a Constructed Wetland Combined with Ultrasound and UV Devices for Wastewater Reuse in Agriculture. *Environ. Technol.* **2013**, *34*, 2327–2336. [CrossRef] [PubMed]

43. Raboni, M.; Gavasci, R.; Urbini, G. UASB Followed by Sub-Surface Horizontal Flow Phytodepuration for the Treatment of the Sewage Generated by a Small Rural Community. *Sustainability* **2014**, *6*, 6998–7012. [CrossRef]

44. Zapater-Pereyra, M.; Lavrnić, S.; van Dien, F.; van Bruggen, J.J.A.; Lens, P.N.L. Constructed Wetroofs: A Novel Approach for the Treatment and Reuse of Domestic Wastewater. *Ecol. Eng.* **2016**, *94*, 545–554. [CrossRef]

45. Melián, J.A.H.; Rodríguez, A.J.M.; Araña, J.; Díaz, O.G.; Henríquez, J.J.G. Hybrid Constructed Wetlands for Wastewater Treatment and Reuse in the Canary Islands. *Ecol. Eng.* **2010**, *36*, 891–899. [CrossRef]

46. Collison, R.S.; Grismer, M.E. Nitrogen and COD Removal from Septic Tank Wastewater in Subsurface Flow Constructed Wetlands: Plants Effects. *Water Environ. Res. Res. Publ. Water Environ. Fed.* **2015**, *87*, 1999–2007. [CrossRef]

47. Carballeira, T.; Ruiz, I.; Soto, M. Effect of Plants and Surface Loading Rate on the Treatment Efficiency of Shallow Subsurface Constructed Wetlands. *Ecol. Eng.* **2016**, *90*, 203–214. [CrossRef]

48. Headley, T.; Nivala, J.; Kassa, K.; Olsson, L.; Wallace, S.; Brix, H.; van Afferden, M.; Müller, R. Escherichia Coli Removal and Internal Dynamics in Subsurface Flow Ecotechnologies: Effects of Design and Plants. *Ecol. Eng.* **2013**, *61*, 564–574. [CrossRef]

49. Nivala, J.; Wallace, S.; Headley, T.; Kassa, K.; Brix, H.; van Afferden, M.; Müller, R. Oxygen Transfer and Consumption in Subsurface Flow Treatment Wetlands. *Ecol. Eng.* **2013**, *61*, 544–554. [CrossRef]

50. Leto, C.; Tuttolomondo, T.; La Bella, S.; Leone, R.; Licata, M. Effects of Plant Species in a Horizontal Subsurface Flow Constructed Wetland—Phytoremediation of Treated Urban Wastewater with *Cyperus alternifolius*, L. and *Typha latifolia*, L. in the West of Sicily (Italy). *Ecol. Eng.* **2013**, *61*, 282–291. [CrossRef]

51. Kasprzyk, M.; Pierzgalski, K.; Wojciechowska, E.; Obarska-pempkowiak, H.; Gajewska, M. Application of Eco-Innovative Technologies of Nutrients Removal in Wastewater—Case Study BARITECH Project. *Rocz. Ochr. Sr.* **2017**, *19*, 143–160.

52. Chen, Y.; Wen, Y.; Zhou, Q.; Huang, J.; Vymazal, J.; Kuschk, P. Sulfate Removal and Sulfur Transformation in Constructed Wetlands: The Roles of Filling Material and Plant Biomass. *Water Res.* **2016**, *102*, 572–581. [CrossRef] [PubMed]
53. Fountoulakis, M.S.; Daskalakis, G.; Papadaki, A.; Kalogerakis, N.; Manios, T. Use of Halophytes in Pilot-Scale Horizontal Flow Constructed Wetland Treating Domestic Wastewater. *Environ. Sci. Pollut. Res.* **2017**, *24*, 16682–16689. [CrossRef]
54. Tian, T.; Tam, N.F.Y.; Zan, Q.; Cheung, S.G.; Shin, P.K.S.; Wong, Y.S.; Zhang, L.; Chen, Z. Performance and Bacterial Community Structure of a 10-Years Old Constructed Mangrove Wetland. *Mar. Pollut. Bull.* **2017**, *124*, 1096–1105. [CrossRef]
55. Arroyo, P.; Blanco, I.; Cortijo, R. Twelve-Year Performance of a Constructed Wetland for Municipal Wastewater Treatment: Water Quality Improvement, Metal Distribution in Wastewater, Sediments, and Vegetation. *Water Air Soil Pollut.* **2013**, *224*. [CrossRef]
56. Garcia-Cuerva, L.; Berglund, E.Z.; Binder, A.R. Public Perceptions of Water Shortages, Conservation Behaviors, and Support for Water Reuse in the U.S. *Resour. Conserv. Recycl.* **2016**, *113*, 106–115. [CrossRef]
57. Zhang, C.; Wen, L.; Wang, Y.; Liu, C.; Zhou, Y.; Lei, G. Can Constructed Wetlands Be Wildlife Refuges? A Review of Their Potential Biodiversity Conservation Value. *Sustainability* **2020**, *12*, 1442. [CrossRef]
58. Ezeah, C.; Reyes, C.A.R.; Gutiérrez, J.F. Constructed Wetland Systems as a Methodology for the Treatment of Wastewater in Bucaramanga Industrial Park. *J. Geosci. Environ. Prot.* **2015**, *3*, 1–14. [CrossRef]
59. Everard, M.; Harrington, R.; McInnes, R.J. Facilitating Implementation of Landscape-Scale Water Management: The Integrated Constructed Wetland Concept. *Ecosyst. Serv.* **2012**, *2*, 27–37. [CrossRef]

Publisher's Note: MDPI stays neutral with regard to jurisdictional claims in published maps and institutional affiliations.

 *sustainability*

Review

A Mini-Review of Urban Wastewater Treatment in Greece: History, Development and Future Challenges

Charikleia Prochaska *  **and Anastasios Zouboulis**

Laboratory of Chemical and Environmental Technology, Department of Chemistry, Aristotle University of Thessaloniki, 54124 Thessaloniki, Greece; zoubouli@chem.auth.gr
* Correspondence: prohaska@chem.auth.gr

Received: 12 July 2020; Accepted: 29 July 2020; Published: 30 July 2020

Abstract: Although Greece has accomplished wastewater infrastructure construction to a large extent, as 91% of the country's population is already connected to urban wastewater treatment plants (WWTPs), many problems still need to be faced. These include the limited reuse of treated wastewater and of the surplus sludge (biosolids) produced, the relative higher energy consumption in the existing rather aged WWTPs infrastructure, and the proper management of failing or inadequately designed septic tank/soil absorption systems, still in use in several (mostly rural) areas, lacking sewerage systems. Moreover, the wastewater treatment sector should be examined in the general framework of sustainable environmental development; therefore, Greece's future challenges in this sector ought to be reconsidered. Thus, the review of Greece's urban wastewater history, even from the ancient times, up to current developments and trends, will be shortly addressed. This study also notes that the remaining challenges should be analyzed in respect to the country's specific needs (e.g., interaction with the extensive tourism sector), as well as to the European Union's relevant framework policies and to the respective international technological trends, aiming to consider the WWTPs not only as sites for the treatment/removal of pollutants to prevent environmental pollution, but also as industrial places where energy is efficiently used (or even produced), resources' content can be potentially recovered and reused (e.g., nutrients, treated water, biosolids), and environmental sustainability is being practiced overall.

Keywords: Greece; wastewater treatment plant (WWTP); history; policy; technology trends and applications

---

1. Introduction

Greece is located in the southern end of the Balkan Peninsula, bordering the Ionian, Aegean, and the Mediterranean Seas. The country has a total area of 131,940 km^2 with a population of (around) 11 million, according to the last official census of 2011. Greece consists of a large mainland and extensive archipelagos of around 3000 islands, has almost 18,000 km of coastline and is 80% mountainous with the highest peak of 2919 m in Olympus [1].

The climate in Greece is predominantly classical Mediterranean (mild, wet winters and hot, dry summers). However, due to the country's geography, a range of micro-climates and local variations can be found [2]. Annual rainfall ranges from 300 to 500 mm in southeastern Greece and from 800 to 1200 mm in the northwestern parts of the mainland, but may exceed 2000 mm in some mountainous areas. Greece is expected to have an 18% precipitation decrease by midcentury, and 22% by the end of the century. Considering an expected modest population decline, Greece's per capita water resources are also expected to decline by midcentury [3,4]. These perspectives are likely to necessitate the implication of appropriate changes to Greece's water resources management and, consequently, changes to the respective ways urban wastewater management is commonly practiced.

Even though Greece is among the very few ancient civilizations, where hygienic technologies and wastewater management were practiced as early as 400 BC [5], the wastewater management history of the country has followed similar development paths as to other nations in Europe, namely Britain, France, and Germany, i.e., for centuries wastewater management was not given much, if any, attention [6]. Although in some ancient cities there were existing sewerage systems, wastewaters were usually disposed of in the streets and near the city centers, creating serious impacts on public health and the environment. The industrialization and urbanization during the second half of the 19th century shifted interest towards better wastewater management practices, while the development of the wastewater management in Greece began around the early 1980s, when the country entered the European Union (EU) and as a member state had to comply with the respective EU policies for proper urban wastewater treatment and thus, created its first urban wastewater treatment plants (WWTPs) [7].

Today, Greece's urban wastewater sector presents significant achievements, but also failures, as well as numerous future challenges. Even though the understanding of the various wastewater treatment processes is important, it is not within this paper's intentions to provide here the in-depth explanations, which can be found elsewhere. This mini-review intends to highlight the major aspects of Greece's urban wastewater treatment history, its recent developments, and present situation, as well as to identify the remaining challenges, in respect to the country's specifications, regarding the European Union's policies and the international technological trends, so as WWTPs to be seen not only as sites for the treatment/removal of pollutants, but also as "industrial" places, where energy can be efficiently used (and produced), and new products (recovery of resources' content) and business opportunities would be created.

We believe the readership of the Journal's Special Issue would be interested in this mini-review, as it highlights not only municipal wastewater management of a particular country, but addresses a country's own wastewater management path in respect to the global thinking of sustainable environmental development.

2. The Past of Urban Wastewater Treatment in Greece

2.1. Historical Development

Historical findings in palaces and cities on the island of Crete, as well as in other Aegean Islands, declare the presence of quite sophisticated sewage systems since the early Minoan civilization, which is considered as Europe's first advanced civilization, flourishing from 2700 to 1450 BC [5]. Sewage systems were found also in residencies in Delos Island (as early as 400 BC) and later, around 300 BC, in the ancient Pella, a city located in northern Greece, best known as the historical capital of the ancient Macedonian kingdom. However, during the same ancient times, open combined sewerage networks of wastewater and storm waters run-off were found also in several other ancient cities, causing quite frequent dispersion of serious water-borne diseases, such as cholera, plague, etc. [8].

This practice has continued for almost the next 15 centuries, before being replaced by the use of absorbing septic tanks. After saturation of the septic tank, either a new tank was built nearby, or the waste was collected and discharged untreated into open streams or into the sea. Athens, the present capital of Greece, was first declared the capital of the newly established Greek State on September 18, 1834. Six years later, the first combined flow sewerage system for the collection of both wastewaters and storm waters run-off was constructed for the first time in the modern history of Greece, noting, however, that Athens at that time was a small city with only 7000 residents.

During the years 1834–1980, the sewerage infrastructure was progressively enlarged in scale. Already in 1950, the preliminary design of the Athens Sewerage System began covering an area of around 200,000 hectares. This design was used as the basis for the development of the city's networks during the 1960s and 1970s. During the same period, similar infrastructures were built also in other large cities of Greece, e.g., in Thessaloniki (the 2nd largest city of Greece), where the first combined sewer overflow pipe was built in 1938, while the basis for the present sewerage system network was set up in 1977 [9].

Entering into the 1980s, the situation started to improve, as Greece joined the EU in 1981. The issue of the treatment of the collected sewage that up to this point was discharged without any treatment to nearby streams, rivers, or to the sea, started to gain particular attention.

2.2. Development over the Past 40 Years

As a member state of the EU, Greece had to comply with the respective EU policies for wastewater treatment. The development of urban WWTPs followed the developments of the relevant EU directives, as well as the available funding from various EU framework programs. Thus, the first modern WWTPs in Greece started their operation around the beginning of 1990s, as a consequence of the respective EU Directive 91/271/EEC [10] "on urban wastewater treatment and disposal" and its amendments, defined by the subsequent Directive 98/15/EEC [11].

The wastewater treatment plant of Thessaloniki started to fully operate in 1992. This plant is situated 7 km southwest of the city, currently serving about 1 million residents of the greater Metropolitan area, by treating daily around 160,000 m^3 with the schedule future capacity (after extension) to be able to treat around 300,000 m^3/day. The treatment process includes screening, grid removal, primary sedimentation (without use of chemical coagulants), conventional activated sludge treatment with nitrogen removal, and effluent disinfection using chlorine gas (Cl_2). The treated effluent is mainly discharged to the nearby Thermaikos gulf, which is characterized as a sensitive area. However, some efforts of effluent reuse in irrigation were started for the first time in Greece in 1993. Approximately 2225 m^3/year of WWTP effluent is being regularly reused, after mixing with freshwater from the nearby Axios River at 1:5 ratio irrigating approximately 2500 ha of spring crops in the nearby Halastra-Kalohori agricultural area [12]. Sewage sludge (i.e., the primary plus the excess activated) is anaerobically digested, thickened, and dewatered and dried (succeeding substantial volume reduction). The greatest amount of this sludge was deposited in an urban landfill, until few years ago when this practice was prohibited, and currently, this is being commonly used as soil amendment after the appropriate control [7].

In 1994, the Psyttalia wastewater treatment plant started its operation. This unit is situated at the uninhabited island of Psyttalia, in the Saronic gulf, between the island of Salamis and the commercial port of Piraeus, approximately 13 km south west of Greece's capital city center (Athens) and serves approximately 4 million of population. At its current state, the wastewater treatment includes pretreatment, primary treatment, and advanced secondary biological treatment, using activated sludge processes with nitrogen removal, thermal sludge drying unit, and co-generation of electricity and heat. The sludge produced is treated anaerobically, resulting in the production of biogas. The produced biogas can cover most of the energy needs of the wastewater treatment facilities. This is one of the largest WWTPs in Europe (and internationally), with a population equivalent (p.e.) coverage of 5,600,000 p.e. The average flowrate of incoming sewage is about 1,000,000 m^3/day. An aerial view of Phyttalia WWTP is presented in Figure 1 [13].

Figure 1. Phyttalia wastewater treatment plant.

The developments of wastewater treatment in the rest of Greece had to follow the requirements and time restrictions as set by the EU Directives 91/271/EEC and 98/15/EEC. According to these, the urban wastewater entering collecting systems should have been subjected to secondary treatment or an equivalent treatment with the following timeline schedule:

1. At the latest by 31 December 1998 for urban agglomerations of more than 10,000 p.e., discharging into receiving waters, which are considered as "sensitive areas", e.g., natural freshwaters bodies; in these cases the relevant WWTPs must be complimentary designed for nitrogen and/or phosphorus removal in order to avoid eutrophication problems of the receiving water bodies.
2. At the latest by 31 December 2000 for all discharges from urban agglomerations of more than 15,000 p.e.
3. At the latest by 31 December 2005 for all discharges from agglomerations of between 10,000 and 15,000 p.e.
4. At the latest by 31 December 2005 for discharges to fresh-water and estuaries water bodies from agglomerations (settlements) between 2000 and 10,000 p.e.

The term "agglomeration" was defined in the Directives as the area where the population and/or economic activities are sufficiently concentrated for urban wastewater to be collected and conducted to a WWTP, or to a final discharge point.

The construction of sewerage networks is currently not legally enforced for agglomerations of less than 2000 p.e., unless the wastewater is discharged to sensitive water bodies. This flexibility in the application of appropriate treatment technologies shifted interest towards the operation of alternative technologies for small communities, but to a limited extent. Sequencing Batch Reactors (SBR) exist in Greece, but are not widely operated. Natural treatment systems, such Waste Stabilization Ponds (WSP) are also quite limited. The first WSP was constructed in 1982 in Sitochori village in the prefecture of Serres in northern Greece, serving a community of 1000 p.e., consisting of a primary facultative pond, followed by two maturation ponds [14]. There was a shift in preference towards the use of constructed wetlands (CWs) in the 1990s, after the operation of the first vertical flow constructed wetland (VFCW) in Nea Madytos community (3000 p.e.) of the Thessaloniki Region in north Greece, back in 1993 [15].

However, the operation of alternative technologies for the wastewater treatment of small communities has not been widely accepted as the feasible alternative to the conventional wastewater treatment systems in Greece. This can be devoted mainly because their application has not being accompanied with a parallel effort to gain wider community acceptance over the conventional wastewater treatment systems, which were already widely applied, broadly tested, and considered as more reliable solutions. Thus, out of 147 WWTPs of small communities in the year 2000, 110 were served by convenient extended aeration systems [16,17].

Figure 2 summarizes the growth of WWTPs in Greece during 1980–2015 [18], while Table 1 shows the information of some milestone WWTPs in the development history of Greece's urban wastewater sector.

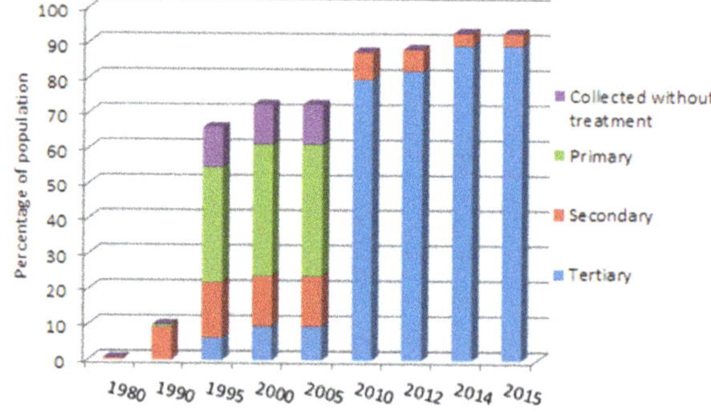

Figure 2. Growth of urban wastewater treatment plants (WWTPs) in Greece during 1980–2015.

Table 1. Milestones of WWTPs development in Greece during 1980–2015.

Year of Operation	WWTP	Capacity (p.e.)	Milestone
1982	Sitochori village (Serres prefecture)	1000	The first WSP, serving a village community [15]
1992	Thessaloniki	1,400,000	The first modern large-scale WWTP in Greece [7]
1993	Nea Madytos (Thessaloniki Prefecture)	3000	The first VFCW serving an urbanity [16]
1992	Thessaloniki	1,400,000	The first WWTP applying reclaimed wastewater reuse projects [12]
1994	Psyttalia	5,600,000	The largest WWTP in Greece and among the largest in Europe [13]
2003	Psyttalia	5,600,000	Operation of the first large-scale biogas unit in WWTP [14]
2007	Psyttalia	5,600,000	The first WWTP operating sludge thermal drying treatment. (Evaporation capacity 34.4 tn/h) [14]
2014	DEYA Heraklion Crete (Finikia)	200,000	The first WWTP applying MBR process [19]

WSP, Waste Stabilization Ponds; VFCW, vertical flow constructed wetland; MBR, Membrane Biological Reactors.

As consequence to the county's alignment with the European Union's Directives and its economic support, Greece presented higher value of percentage of population connected to tertiary wastewater treatment in the year 2015, 88% as compared to 80.1% of mean value for central European countries (Austria, Belgium, Denmark, Netherlands, Germany, Switzerland, Luxembourg, and United Kingdom) [18], which are considered more technologically advanced in the field of wastewater treatment and economically stronger than Greece (countries economic status is a key aspect closely related to wastewater treatment status [20]). Meanwhile, the international mean value of percentage of population connected to tertiary wastewater treatment in the year 2015 was 59% (calculated mean of data from 27 countries) [21].

With respect to Greece's milestones, presented in Table 1, Greece presented a delay between applying the appropriate technology and the technology's first application elsewhere, justified by the social-political-economic constraints that needed to be overcome, institutional arrangements, research and development, and technical and economic co-operation that needed to be established first. In North America, natural treatment systems, treating municipal wastewater, started at the end of the 1960s and beginning of the 1970s [22]. In Sweden, biogas has been produced at municipal wastewater treatment plants since the 1960s [23]. Tertiary treatment of municipal wastewater was first introduced in 1962 in the USA by Lutdzack and Ettinger [24]. In the late 1980s, manufacturers who had successfully applied sludge thermal drying technologies in chemical and food industries, transferred existing technologies to sewage sludge [25]. The first such applications are reported in Avonmouth and Countess Wear in England [26], while by 1993 Membrane Biological Reactors (MBR) systems had been reported for use in sanitary application in Europe [27].

3. Recent Developments and Current Situation

Today, Greece has a population of approximately 11 million inhabitants, according to the last official census of 2011. The main management model for water and wastewater services in Greece is delegated public management, although more recently some public-private management schemes have been also started to be implemented. In the two biggest cities of Greece (Athens and Thessaloniki), the Greek state delegated two public companies (by a majority) named E.YD.A.P for Athens and E.Y.A.TH. for Thessaloniki. These companies have been listed also in the Greek stock market already for more than 15 years and the Greek state is the main shareholder and, therefore, the President of the Board of Directors and the CEO of both companies are appointed by the government. For cities with more than 10,000 inhabitants, the national law regulating the role and the foundation of Municipal Water and Sewerage Companies was published in 1980. Currently, there are 130 municipality-owned companies, named D.E.Y.A. In cities with fewer than 10,000 inhabitants, the management model is usually the direct urbanity public management [28].

According to article 17 of the Urban Wastewater Treatment Directive 91/271/EEC [10], each member state has to prepare every two years the respective Implementation Report, describing the collection, treatment, and disposal of the relevant wastewater data. Based on the most recent report of 2017, which refers to the year 2014, 91% of the country's population is connected to 254 urban wastewater treatment plants, treating almost 1.74×10^6 m^3/day. These plants provide secondary biological treatment, with 83% providing biological nitrogen removal, 57% providing (additionally) biological phosphorus removal, and 93% providing (tertiary) wastewater disinfection [29].

In 2012, a National Database was created by the Specific Secretariat for Water (SSW) of the Greek Ministry of Environment and Energy, with the initiative and the support of the European Commission Directorate-General for Regional and Urban Policy. This Database has been upgraded and enriched ever since, presenting all the relevant information content in an advanced Geographical Information System (GIS), easily accessed by everyone interested via the respective webpage ("SSW—Wastewater Treatment Plants") [30]. Within the Database, specific information regarding the location, capacity, performance, means of disposal, or reuse of wastewater and of sludge, as well as the Environmental Terms of each WWTP are stored and are easily publicly accessible. Figure 3 represents the map of Greece's WWTPs, as presented on SSW webpage, with the respective locations. In this figure there are WWTPs in compliance with the requirements of the Directive 91/271/EEC, as indicated by blue dots (which are representing the majority, i.e., 168 out of 254). The figure presents the WWTPs that are not still compliant with the Directive, either because they do not collect a sufficient number of samples per year, or because the effluent is outside of the respective limits as set by the Directive, which are indicated by red in color dots. Each dot size is logarithmically related to the capacity of each WWTP [31].

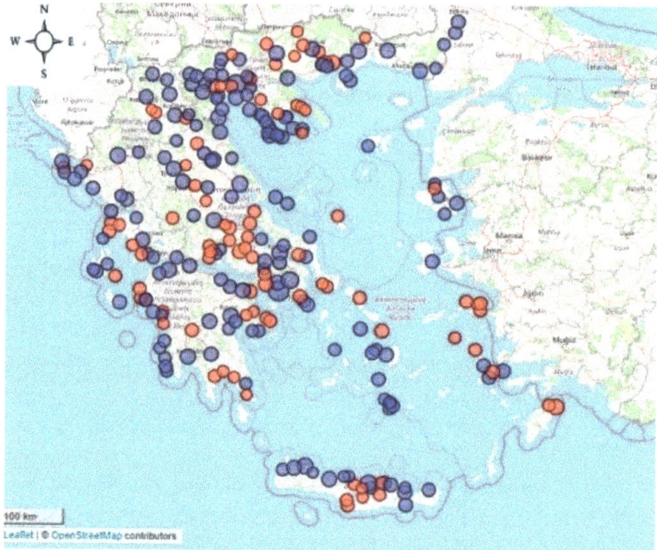

Figure 3. Map of Greece's WWTPs.

Despite the aforementioned developments, EU Commission has brought Greece several times to the EU Court of Justice for violating Directive 91/271/EEC requirements, regarding specific cases. The first condemn took place in 2004 [32], and the Court ruled that Greece was violating EU law by not adequately collecting and treating wastewater discharged into the Gulf of Elefsis, which is characterized as a sensitive area. Twelve years later, the judges ruled that Greece continued to be in violation of the previous 2004 ruling, by not taking the "measures necessary for the installation of a collecting system for the urban wastewaters from the Thriasio Plain, in western Attica and not subjecting the urban wastewater from that area to more stringent treatment than the secondary one, before discharged into the sensitive area of the Gulf of Elefsis", imposing a fine of EUR 5 million, with an additional penalty of EUR 3.28 million for every six months the government fails to meet the required wastewater treatment regulations [33].

In 2007, the Court found that Greece had failed to fulfil its obligations on the grounds that 23 urban agglomerations of over 15,000 p.e. were still not equipped with the appropriate systems for collecting and/or treating the urban wastewaters [34]. In 2014, the court found that Greece had still not complied with the previous 2007 judgment in 6 out of the 23 agglomerations, imposing a fixed sum of EUR 10 million and a sliding-scale periodic fine of EUR 20,000 €/day [35]. For the aforementioned fines Greece had to pay EUR 10,558,145 in the year 2019.

Regarding the urban agglomerations between 2000 and 10,000 p.e., the main issues that Greece has to phase is that of the 385 agglomerations belonging in this category, as there are only 123 in full compliance with the requirements of the Directive 91/271/EEC. Another 162 have included or are in the current projects of the National Strategic Reference Framework (NSRF). If these projects are implemented and functioning properly (noting that neither of them is obvious), then in few years the coverage rate will rise to 74% from today's 32%. The rest of these settlements have some infrastructure missing, have several operational problems, or have incomplete measurements taken and, therefore, they are not shown to meet the rather strict Directive requirements. In order to promote the respective wastewater infrastructures of this category, Greece, with the encouragement of the European Commission, set up in 2018 a committee to design and implement a united national wastewater infrastructure master plan, consisting of 13 regional plans, one for each first-level administrative region of the country. Results of this initiative will soon be published [36].

Agglomerations of less than 2000 inhabitants are accounting for almost 2.5 million p.e. in Greece; in these cases, neither sewage networks nor wastewater treatment are legally enforced [37]. However, there are several natural systems in operation for the treatment of wastewater from small communities, although there is no formal registry of these systems and a detailed reference is quite difficult to be provided, as the relevant references are mainly based on literature review. Waste Stabilization Ponds (WSP) are not so popular in Greece. It is estimated that approximately 20 such systems have been constructed in Greece, serving either single small communities, or two to three neighboring communities being in the range of 600–3000 p.e. The majority of these systems exist in two Prefectures of North Greece (Kavala and Serres). All these systems consist of a primary facultative pond, followed by one to three maturation ponds. Nevertheless, most of these ponds present several operation and maintenance problems, such as odors, rooted plants, etc. [15].

Not more than 20 are the literature-referenced CWs that are currently in operation in Greece, with a capacity between 8–1300 p.e. The majority of these systems consist of one to three vertical subsurface flow (VSSF) beds for primary treatment, one to two VSSF beds for secondary treatment, and one to two beds of horizontal subsurface flow (HSSF) for tertiary treatment; disinfection is performed either by means of UV or chlorination. However, CWs systems that use only free water surface systems (FWS) or HSSF in their configuration are also in operation [38,39].

For the majority of wastewaters that originate from small villages, as well as from the lots of decentralized holiday residences in Greece, septic tanks/soil absorption systems remain as the predominant option for sewage treatment. However, the exact number of these systems is still unknown, since they are seldom formally registered.

4. Outlook

4.1. Remaining Gaps and Future Challenges

Greece's wastewater sector has achieved undoubtable progress in terms of infrastructure, technology, and policy implementation, providing a solid ground for its future development. However, in order to be sustainable in this future, providing that wastewater quantity will continue to increase further and that water shortages will become more common [40], specific focus should be provided to the following main objectives:

- Treated wastewater reuse
- Sludge disposal
- Energy efficiency
- Reducing greenhouse emissions
- Management of wastewaters from small settlements that still use quite extensively the septic tanks option

4.1.1. Treated Wastewater Reuse

In a 2018 report, the EU Joint Research Center (JRC) projected an increased pressure on water resources of 20% or more by 2050, as compared to 2010 in Greece [40]. As all southern European countries are expected to face decreasing water availability, the EU initiated the development of a proposal for the regulation of minimum quality requirements for (treated) wastewater reuse, mainly for agricultural irrigation and/or in the industrial sector. The regulation that is expected to be published in the Official Journal of the EU during 2020 will enter into force on the 20th day after its publication and shall apply to the member states three years after the date of its entry into force. According to estimations, the proposed regulation could lead to a substantial increase of water reuse in agricultural irrigation up to 6.6 billion m^3/year across EU member states, as compared to estimated current 1.7 billion m^3/year in the absence of any EU legal framework [41].

This regulation proposes less strict microbial standards in terms of *E. coli*, than the maximum value of 200 cfu/100 mL that is currently legally enforced in Greece for unrestricted irrigation (i.e., applicable to all kind of crops, independent from the irrigation method, where public access may also be allowed). The regulation has no provisions for the maximum permissible concentrations of selected heavy metals and metalloids, as well as for certain agronomic characteristics of the reclaimed water for agricultural irrigation, provided by the Greek relevant regulations [42,43]. In that sense, it might set a new perspective on the issue of reclaimed water reuse in the country.

However, it lacks to address other issues that concern Greece's reclaimed water reuse, such as transportation expenses, due to the fact that most WWTPs are located quite far away from arable agricultural land, which poses another economical challenge for the reclaimed wastewater management in Greece [44]. This explains why the reclaimed water reuse for agricultural irrigation is practiced in only 13% of the existing WWTPs in Greece. The main WWTP of Athens, for instance, located on the Psytalia island, reuses onsite part of its effluent that undergoes filtration (through sand-filters) and disinfection (by means of UV devices), so as to be reused as process water for the treatment facilities [14]. Meanwhile, it is estimated that almost 18,000 ha are being irrigated by the several agricultural water reuse projects in Greece, whereas almost 60,000 more ha are irrigated via the indirect wastewater reuse. As Greece's irrigated land sums up to almost 103,860,000 ha/year [45], there is a huge potential for turning to reclaimed wastewater, rather than using freshwater to cover the needs of the agricultural land, providing that all the necessary concerns regarding human health and environmental protection are met.

4.1.2. Sludge Disposal

Based on the most recent published data, the disposal situation of wastewater sludge from Greece's WWTPs is presented in Figure 4 [46]. In 2016, the sludge production was at the level of 119,770 tones, with the majority of it (53%) ending up in landfills, 33% used in agriculture, and only 18% used for composting or other specific applications, such as alternative fuel in cement industries.

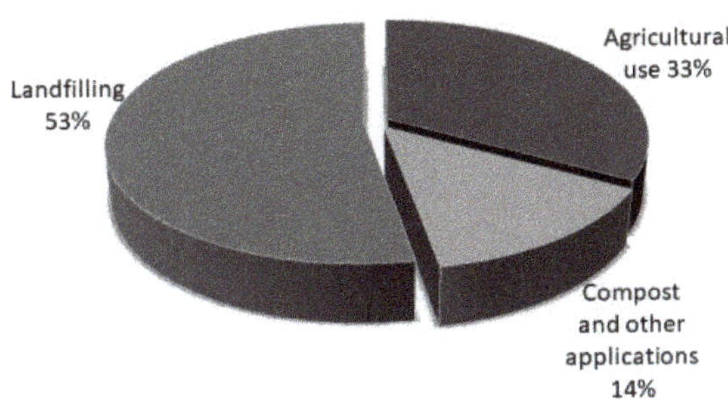

Figure 4. Disposal situation of wastewater sludge from Greece's WWTPs in 2016.

Even though Directive 86/278/EEC, the so-called Sewage Sludge Directive [47], was adopted in Greek legislation back in 1986, aiming to encourage the application of sewage sludge in agriculture and to regulate appropriately its use, attempting to prevent any harmful environmental and human effects, nevertheless the current use of WWTPs sludge for land application is rather limited. This is mainly due to the fact that sludge produced through successive anaerobic digestion/stabilization, dewatering, and thermal drying in the megacities (Athens, Thessaloniki) has a fairly higher heavy metal content [48], than the thresholds of the Ministerial Decision 80568/4225/91 [49], which set supplementary national thresholds for the chromium species Cr(III) (500 mg kg^{-1}) and Cr(VI) (10 mg kg^{-1}), apart from the limits of other metal contents as stipulated by Directive 86/278/EEC [50]. For this reason, the dried sludge from the bigger treatment plants is potentially used as a fuel substitute in the cement industry, but also (partly) as soil conditioner.

According to the National Plan on Waste Management (CMA 49/15.12.2015/GG174A) [51] and in accordance to the Directive 2008/98 on waste [43]), the sludge disposal to landfill should be minimized to 5% by 2020, while the recovery should be 95% (energy recovery and agricultural use). In this context, few WWTPs have invested also in Bio-augmentation technology projects that may reduce substantially the sludge, produced in the treatment facilities, by 50%–80% [52]. As such projects are quite limited, there is still scope in Greece for further alignment in order to achieve a sustainable sludge disposal strategy.

4.1.3. Energy Efficiency

Energy is a key factor in achieving sustainability in the WWTPs sector, and a shift from the negative energy balance (i.e., when energy demand is being covered by external sources) to the energy neutral, or even to energy positive, wastewater treatment has been postulated worldwide [53].

Biogas produced at Psyttalia WWTP is being utilized onsite as the fuel in two Combined Heat and Power (CHP) plants, of 11.4 MW power capacity. The CHP plant system provides a considerable part of the heat needs of Psyttalia WWTP (especially for sludge digestion and drying), as well as for the electric power needs (e.g., for aeration in biological treatment step), whereas the surplus power is being sold to the National Power Grid Manager [14]. Meanwhile, the plant continues to participate in innovative projects that would further enhance its energy efficiency. The SMARTech4b pilot scale system (part of the ongoing SMART-Plant project, funded under European Union's Horizon 2020 research and innovation program) will be tested and validated at the WWTP of Psyttalia, aiming to enable the integration of enhanced biogas recovery (by preliminary thermal hydrolysis) of sewage sludge with side-stream, energy-efficient, and compact nitrogen removal and phosphorus recovery, avoiding the problematic struvite formation [54].

Meanwhile, other research programs are ongoing at smaller WWTPs in Greece. e.g., the LIFE B2E4 sustainable-WWTP project (funded by the European Commission under the LIFE Framework Programme) [55] aims to improve the performance of overloaded extended aeration wastewater treatment plants, by using a novel process for removing solids prior to aeration. For this purpose, a micro-screening system for the removal of oversized solids will be installed at DEYA's WWTP, serving the Rethymno area of Crete Island (located in the southern part of Greece). The plant, which applies an extended aeration activated sludge process, has an average daily flow of 13,000 to 15,000 m^3. Its peak flow capacity is about 17,000 m^3/day. In addition, the project will demonstrate the valorization of produced biosolids for the production of electric energy (mainly through gasification), thus reducing further the net energy consumption of the WWTP. The produced electric energy will be sufficient for energy self-sustainable operation for wastewater treatment. According to theoretical estimations, waste heat from the plant itself can provide energy to 10% of the buildings that deliver wastewater to the WWTP [56].

Even though there is no specific national or EU legislation to be followed for improving the overall energy performance of WWTPs, with the majority of plants built more than 20 years ago in Greece, the opportunity of promoting the energy efficient wastewater resource recovery concept—through technological improvements within the existing plants to transform the wastewater treatment plants into real power producers (i.e., prosumers, instead of consumers) and to eventually prompt the development of new products and business opportunities—is another challenge for the wastewater treatment sector that Greece cannot overlook.

4.1.4. Reducing Greenhouse Emissions

Improving energy efficiency in urban wastewater facilities can help reduce Green House Gas (GHG) emissions from the WWTPs, where GHG are largely in the form of carbon dioxide (CO_2), methane (CH_4), and nitrous oxide (N_2O). CO_2 is mainly produced during the applied wastewater and sludge treatment processes, as well as through the consumption of energy during WWTPs' operation [57]. CH_4 is mostly emitted during the sludge treatment works (e.g., sludge anaerobic digestion, sludge disposal to landfills), while it can also leak from the biogas plants (due to poor design or maintenance), and N_2O can be emitted during the biological processes, aiming toward nitrogen removal [58]. CO_2 is not accounted as GHG from WWTPs' operation, due to its biogenic origin, whereas the main focus is directed to CH_4 and N_2O gases, which present a global warming potential 28 and 265 times higher, respectively, than the CO_2, considering a 100-year period [59].

Greece, as an EU member state, has committed to a 40% GHG emissions' reduction by 2030, as compared to 1990 levels. Because GHG emissions from the urban wastewater treatment (0.458 million tons of the specific CO_2 equivalent (CO_2e) in 2017, according to Eurostat data, [60]) account for approximately one-fifth of the emissions from the overall waste sector, it remains important to carefully manage CH_4 and N_2O gases and to accomplish the overall GHG emissions targets.

The quantification of GHG emissions from WWTPs in Greece is rather a difficult task, due to unavailable formal registered data and quantifying the models' approach complexity. A recent study reported that the two major WWTPs of Greece (Athens and Thessaloniki) contribute to 67.1% of the total emitted GHG, as expressed in tones/day of CO_2e. The contribution of WWTPs with capacity ranging between 10,000 and 100,000 inhabitants accounts for 20.6%, while the contribution of the other groups of WWTPs is less than 10%, either due to their small number, or due to the lower volumes of wastewaters, which are treated daily [61].

However, as reported in the literature, GHG emissions depend mainly on the wastewater composition and the applicable treatment technologies and, hence, they are not always linked to the respective treatment plants' capacity (actual loads) [54]. Thus, the focus is devoted to the minimization and prevention rather than the treatment of GHG emissions in WWTPs, as up to now most of the technologies available for the treatment of GHG are quite expensive, or even not suitable/effective enough to treat the gaseous streams of WWTPs [62].

Thus, when choosing the configuration of new WWTPs and deciding on how to upgrade the existing (aged) ones, the proper evaluation of the treatment processes should be performed, attempting to lower the contribution towards global warming, being as little as possible. The application of certain models, estimating GHG emissions, could be adopted by the National Database of the SSW (Specific Secretariat for Water) of the Greek Ministry of Environment and Energy, so as to help the operators of treatment systems to reduce GHG emissions from the existing WWTPs. Finally, to enhance this target, a limit value of CO_2e emission/m^3 of treated wastewater, e.g., <2 kg CO_2e/m^3, as recommended by Koutsou et al. [61] (compared to 2.2 kg CO_2e/m^3 of currently estimated for Greece [47] and to 0.8–1 kg CO_2e/m^3 of currently estimated for anaerobic/anoxic/oxic treatment processes for urban WWTPs [62]), could be adopted in the national legislation. Note also that the limit value does not have to remain constant over time, but it should be regularly updated in respect to the observed global trends, regarding the GHG emissions from the WWTP sector.

4.1.5. Management of Wastewater from Small Urban Settlements that Still Use Septic Tanks

In Greece, there are many agglomerations with less than 2000 inhabitants, which account for almost 2.5 million p.e., where sewerage networks are not legally enforced and where the septic tank/soil absorption systems are the common norm [37].

The Special Secretariat for Water of the Greek Ministry of Environment published in 2012 specific guidelines regarding the wastewater management from small settlements (SSW) [37]. This text provided directions for the onsite wastewater treatment alternatives to the septic tank/soil absorption systems, by codifying the international experience of the existing till then technology making the appropriate adaptations to the country's particularities. Since then, the alternative onsite wastewater treatment technologies and the relevant research have evolved in Greece, e.g., regarding the use of CWs [38], Sequencing Batch Reactors (SBR) [63], or even the Membrane Biological Reactors (MBR systems) [64], and other combinations of treatment techniques, such as MBR-RO (Reverse Osmosis) [65], following the respective international trends on adapting these technologies for the onsite wastewater management and reuse in small settlements [66], as well as for the tourist-based communities [67]. However, neither the existing guidelines of SSW have been revised, nor have the alternative technologies of wastewater management, penetrated in any large extend to solve the problem of not-sewerage areas of Greece, leaving the respective situation practically unchanged.

There is always a concern with respect to the adverse environmental impacts and local public health risks of failing, or inadequately designed septic tank/soil absorption systems, used in the not-sewerage areas of Greece [68]. As such, systems receive an increased load during the holiday seasons (particular in summer, where the population in several small settlements is doubled or even more) and due to the rapid growth of short-term holiday rentals of such type of possible remote residences, especially over the last years [69], and also due to an increasing number of refugees and migrants (especially for certain Aegean islands) [70,71]. The systems may then face water shortage problems [72,73], and where the number of the hosted migrants may raise the native population by one-third or even more [70]. There is an urge that management of wastewater from the small settlements should not be left out of policies, regarding wastewater treatment and wastewater reuse in general, especially in the water-scarce areas of the country. In fact, the public acceptance of reused wastewater as a resource may be gained, if specific attention is driven first to the proper management and reuse of wastewater, produced at such small scale.

A policy reform towards this direction could include: (a) *a mandatory registry* of each type of wastewater system that is not connected to centralized sewage treatment plants, aiming this registry to serve as a dynamic tool for the authorities for providing directions, instructions, and support, regarding the proper wastewater treatment and reuse, thus upgrading the role of SSW's guidelines; (b) support funding for the efficient (residential and small business, e.g., hotels) decentralized wastewater management and reuse projects.

As currently the EU is searching effective ways to emerge from the COVID-19 crisis by injecting the economy with funds [74], the ecological part and the greening of the economy, regarding the onsite decentralized treatment level of wastewaters, should not be overlooked. In fact, a promising area of the ongoing research on COVID-19 involves using sewage to monitor virus circulation in communities and to detect possible outbreaks, even before clinical cases have been identified [75]. In that sense, as novel enveloped viruses are expected to emerge, when leaving homes or staying at decentralized hotels, where proper onsite wastewater management and reuse is applied, this approach could help tackle, at the local scale, and avoid larger-scale virus outbreaks in the future.

The tools and experience gained by e.g., the relevant Greek-EU co-funded "Saving Energy at Home" Programme, a grant and loan program that recently run again in Greece [76], could also help in planning of a relevant program that would ease owners to select and implement the adequate wastewater treatment and reuse technology, without being discouraged by the amount of work, the administrative complexity, and certain technical challenges.

5. Conclusions

When reviewing the development history of Greece's wastewater sector, there are both achievements and failures. Even though the country has accomplished its wastewater infrastructure construction to a large extent, many problems are still existing. These include the reuse of treated wastewater and the disposal of produced excess sludge, the remaining high energy consumption of the existing rather aged WWTPs, and the environmental impacts and local public health risks of failing or inadequately designed septic tank/soil absorption systems, still used in the not-sewerage areas of Greece (small settlements, located in mountainous areas or islands), with agglomerations of less than 2000 inhabitants, where sewage networks are not legally enforced.

Looking forward, there will be more challenges in the future, due to the multiple pressures of environment protection, water shortage, economic development (e.g., tourism), or economic crisis. To address these challenges, specific care should be directed to the country's own characteristics and needs, as well as to including global environmental thinking and international technological trends in the development of new policies and the operation of WWTPs, in ways that would promote both resource recovery and environmental sustainability.

Author Contributions: Writing—original draft, C.P.; writing—review and editing, A.Z. All authors have read and agreed to the published version of the manuscript.

Funding: This research received no external funding.

Conflicts of Interest: The authors declare no conflicts of interest.

References

1. EEA. European Environment Agency: Surface Water Quality Monitoring—Summary: Greece. Available online: https://www.eea.europa.eu/publications/92-9167-001-4/page010.html (accessed on 2 May 2020).
2. Philandras, C.M.; Nastos, P.T.; Repapis, C.C. Air temperature variability and trends over Greece. *Glob. NEST J.* **2008**, *10*, 273–285.
3. Chenoweth, J.; Hadjinicolaou, P.; Bruggeman, A.; Lelieveld, J.; Levin, Z.; Lange, M.A.; Xoplaki, E.; Hadjikakou, M. Impact of climate change on the water resources of the eastern Mediterranean and Middle East region: Modeled 21st century changes and implications. *Water Resour. Res.* **2011**, *47*, W06506. [CrossRef]
4. Ferronato, N.; Torretta, V. Waste Mismanagement in Developing Countries: A Review of Global Issues. *Int. J. Environ. Res. Public Health* **2019**, *16*, 1060. [CrossRef] [PubMed]

5. Angelakis, A.N.; Spyridakis, S.V. The status of water resources in Minoan times—A preliminary study. In *Diachronic Climatic Impacts on Water Resources with Emphasis on Mediterranean Region*; Angelakis, A.N., Issar, A., Eds.; NATO ASI Series; Springer: Heidelberg, Germany, 1996; Volume 36, pp. 161–191.

6. Lofrano, G.; Brown, J. Wastewater management through the ages: A history of mankind. *Sci. Total. Environ.* **2010**, *408*, 5254–5264. [CrossRef] [PubMed]

7. Andreadakis, A.; Agelakis, T.; Adraktas, D. Treatment and disposal of the waste-water of Thessaloniki, Greece. *Environ. Int.* **1993**, *19*, 291–299. [CrossRef]

8. Golfinopoulos, A. The Diachronic Management of Waste in Antiquity in Greece. Bachelor's Thesis, Hellenic Open University, Patra, Greese, 15 March 2016.

9. Historic Overview of Water Supply. Available online: https://www.eydap.gr/en/TheCompany/Water/HistoricalTrackBack/ (accessed on 21 March 2020).

10. Directive 91/271/EEC of 21 May 1991 on Urban Waste-Water Treatment 91/271/EEC. Official Journal of the European Communities, Brussels. 1991. 13p. Available online: https://eur-ex.europa.eu/legalcontent/EN/TXT/PDF/?uri=CELEX:31991L0271&from=EN (accessed on 21 March 2020).

11. Directive 98/15/EEC of 27 February 1998. Available online: https://eur-lex.europa.eu/LexUriServ/LexUriServ.do?uri=CONSLEG:1998L0015:19980327:EN:PDF (accessed on 21 March 2020).

12. Ilias, A.; Panoras, A.; Angelakis, A. Wastewater Recycling in Greece: The Case of Thessaloniki. *Sustainability* **2014**, *6*, 2876–2892. [CrossRef]

13. E.YD.A.P Wastewater Treatment. Available online: https://www.eydap.gr/TheCompany/DrainageAndSewerage/Sewerage/ (accessed on 21 March 2020).

14. E.YD.A.P Psyttalia Wastewater Treatment Plant. Available online: https://www.eydap.gr/userfiles/c3c4382d-a658-4d79-b9e2-ecff7ddd9b76/Fact-sheet-PWWTP.pdf (accessed on 21 March 2020).

15. Kotsovinos, N.; Tsagarakis, K.P.; Tsakiris, K.; Agricultural, N. Waste stabilization ponds in Greece: Case studies and perspectives. In Proceedings of the 6th International Conference on Waste Stabilisation Ponds, Avignon, France, 28 September–1 October 2004; IWA: London, UK.

16. Gikas, G.D.; Tsihrintzis, V.A. Urban wastewater treatment using constructed wetlands. *Water Util. J.* **2014**, *8*, 57–65.

17. Tsagarakis, K.P.; Mara, D.D.; Horan, N.J.; Angelakis, A.N. Small urban wastewater treatment plants in Greece. *Water Sci. Technol.* **2000**, *41*, 41–48. [CrossRef]

18. Changes in Urban Waste Water Treatment in Southern Europe. Available online: https://www.eea.europa.eu/data-and-maps/daviz/changes-in-wastewater-treatment-in-10#tab-dashboard-01 (accessed on 21 March 2020).

19. Wastewater Treatment Plant in Finikia is in Operation. Available online: https://www.heraklion.gr/municipality/press-releases-2014/se-leitourgia-h-monada-epexergasias-apovlhtwn-22082014.html (accessed on 10 July 2020). (In Greek)

20. Sato, T.; Qadir, M.; Yamamoto, S.; Endo, T.; Zahoor, A. Global, regional, and country level need for data on wastewater generation, treatment, and use. *Agric. Water Manag.* **2013**, *130*, 1–13. [CrossRef]

21. Wastewater Treatment (%) Population Connected. *OECD Srat.* Available online: https://stats.oecd.org/Index.aspx?DataSetCode=WATER_TREAT (accessed on 27 July 2020).

22. Naja, G.; Volesky, B. Constructed Wetlands for Water Treatment. *Compr. Biotechnol.* **2011**, *6*, 353–369. [CrossRef]

23. Eriksson, O.; Bisaillon, M.; Haraldsson, M.; Sundberg, J. Improvements in Environmental Performance of Biogas Production from Municipal Solid Waste and Sewage Sludge. In Proceedings of the World Renewable Energy Congress, Linköping, Sweden, 8–13 May 2011; Linkoping University Electronic Press: Linköping, Sweden, 2011; Volume 57, pp. 3193–3200.

24. Ludzack, F.J.; Ettinger, M.B. Controlling operation to minimize activated sludge effluent nitrogen. *J. Water Pollut. Control Fed.* **1962**, *34*, 920–931.

25. Chen, G.; Yue, P.L.; Mujumdar, A.S. Sludge Dewatering and Drying. *Dry. Technol.* **2002**, *20*, 883–916. [CrossRef]

26. Hudson, J.A. Treatment and Disposal of Sewage Sludge in the Mid-1990s. *Water Environ. J.* **1995**, *9*, 93–100. [CrossRef]

27. Aya, H. Modular membranes for self contained reuse systems. *Water Qual. Int.* **1994**, *21*, 128–136.

28. Renou, Y. The Governance of Water Services in Developing Countries: An Analysis in Terms of Action Stratification. *J. Econ. Issues* **2010**, *44*, 113–138. [CrossRef]

29. European Commission. *Ninth Report on the Implementation Status and the Programmes for Implementation (As Required by Article 17) of Council Directive 91/271/EEC Concerning Urban Waste Water Treatment*; Report: COM (2017) 749 Final; European Commission: Brussels, Belgium, 2018; 18p.

30. SSW. Wastewater Treatment Plants. Available online: http://astikalimata.ypeka.gr/Services/Pages/WtpViewApp.aspx (accessed on 21 March 2020).

31. Map of Greece's WWTPs. Available online: http://astikalimata.ypeka.gr/Services/Pages/WtpViewApp.aspx (accessed on 21 March 2020).

32. EU Court of Justice. Judgement No. C-119/02 of 24 June 2004. Available online: https://op.europa.eu/en/publication-detail/-/publication/5f3eaad5-3180-423a-806c-f091d7a5ba60/language-en (accessed on 21 March 2020).

33. EU Court of Justice. Judgement No. C-328/16 of 22 February 2018. Available online: https://op.europa.eu/en/publication-detail/-/publication/09a5d1a1-4149-11e8-b5fe-01aa75ed71a1/language-en/format-PDF/source-120866608 (accessed on 21 March 2020).

34. EU Court of Justice. Judgement No. C-440/06 of 25 October 2007. Available online: https://op.europa.eu/en/publication-detail/-/publication/46d1f9fc-c804-4301-811c-e70e5d62dbd7/language-en (accessed on 21 March 2020).

35. EU Court of Justice. Judgement No. C-167/14 of 15 October 2015. Available online: https://op.europa.eu/en/publication-detail/-/publication/ddfaac60-9cb3-11e5-b792-01aa75ed71a1/language-en/format-PDF (accessed on 21 March 2020).

36. Lalios, G. Fewer than Half the Small Greek Cities with Sewage Network. Available online: https://www.kathimerini.gr/1017094/article/epikairothta/ellada/ligoteres-apo-tis-mises-oi-mikres-ellhnikes-poleis-me-apoxeteysh (accessed on 21 March 2020).

37. SWW. Guidelines Text for Wastewater Management of Small Settlements. Available online: http://www.ypeka.gr/LinkClick.aspx?fileticket=0yUWZWCWi3s%3d&tabid=251&language=el-GR (accessed on 21 March 2020).

38. Lavrnić, S.; Mancini, M.; Lavrni, S. Can constructed wetlands treat wastewater for reuse in agriculture? Review of guidelines and examples in South Europe. *Water Sci. Technol.* **2016**, *73*, 2616–2626. [CrossRef]

39. Zouboulis, A.; Prochaska, C.; Angelakis, A. Constructed wetlands wastewater treatment plants and agricultural reuse status in Greece. Presented at the ISCW 2019, Patras, Greece, 29–31 August 2019.

40. Bisselink, B.; Bernhard, J.; Gelati, E.; Adamovic, M.; Guenther, S.; Mentaschi, L.; De Roo, A. *Impact of a Changing Climate, Land Use, and Water Usage on Europe's Water Resources*; EUR 29130 EN; Publications Office of the European Union: Luxembourg, 2018.

41. European Commission. *Proposal for a Regulation of the European Parliament and of the Council on Minimum Requirements for Water Reuse*; Report: COM (201)337/975937; European Commission: Brussels, Belgium, 2018; 28p.

42. CMD (Common Ministerial Decision). *Measures, Limits and Procedures for Reuse of Treated Wastewater*; Number 145116; Energy and Climate Change; Ministry of Environment: Athens, Greece, 2011. (In Greek)

43. CMD (Common Ministerial Decision). *Amendment of CMD 145116 Measures, Limits and Procedures for Reuse of Treated Wastewater*; Number 191002; Energy and Climate Change; Ministry of Environment: Athens, Greece, 2013. (In Greek)

44. Angelakis, A.N.; Asano, T.; Bahri, A.; Jiménez, B.E.; Tchobanoglous, G. Water Reuse: From Ancient to Modern Times and the Future. *Front. Environ. Sci.* **2018**, *6*, 1–17. [CrossRef]

45. Statistics—ELSTAT. B03. Holdings with Irrigable and Irrigated Areas, by Region and Department Excel File. Available online: http://www.statistics.gr/en/statistics/-/publication/SPG31 (accessed on 21 March 2020).

46. Statistics—EUROSTAT. Sewage Sludge Production and Disposal from Urban Wastewater (in Dry Substance (d.s)). Available online: https://ec.europa.eu/eurostat/databrowser/view/ten00030/default/table?lang=en (accessed on 21 March 2020).

47. Directive 86/278/EEC of the 4 July 1986. On the Protection of the Environment and in Particular of the Soil When Sewage Sludge is Used in Agriculture. Available online: https://eur-lex.europa.eu/legal-content/EN/TXT/?uri=celex%3A31986L0278 (accessed on 21 March 2020).

48. Spanos, T.; Ene, A.; Patronidou, C.S.; Xatzixristou, C. Temporal variability of sewage sludge heavy metal content from Greek wastewater treatment plants. *Ecol. Chem. Eng. S* **2016**, *23*, 271–283. [CrossRef]

49. *Ministerial Decision: Greek Legislation 80568/4225/91 of 22 March 1991 on Methods, Specifications and Requirements for the Use in Agriculture of the Sludge Originating from Household and Urban Waste Treatment, Harmonized from European Council Directive*; 86/278/EEC; Official Government Gazette of the Hellenic Republic: Athens, Greece, 1991.

50. *CMA (Council of Ministers Act) 49 of 15-12-2015 (Government Gazette 174/A/15-12-2015)*; Official Government Gazette of the Hellenic Republic: Athens, Greece, 2015. (In Greek)

51. Directive 2008/98/EC of 19 November 2008. On Waste. Available online: https://eur-lex.europa.eu/legal-content/EN/TXT/?uri=CELEX:32008L0098 (accessed on 21 March 2020).

52. Bio-Augmentation for the Sludge Reduction in the WWTP of Patras, Greece. Available online: https://alpha-bioenergy.com/project/bio-augmentation-for-the-sludge-reduction-in-the-wwtp-of-patras-greece/ (accessed on 26 March 2020).

53. Gao, H.; Scherson, Y.D.; Wells, G.F. Towards energy neutral wastewater treatment: Methodology and state of the art. *Environ. Sci. Process. Impacts* **2014**, *16*, 1223–1246. [CrossRef]

54. SMART-Plant Project. European Union's Horizon 2020 Grant No. 690323. Available online: https://www.smart-plant.eu/index.php/psyttalia (accessed on 26 March 2020).

55. Biosolids2Energy Research Project (LIFE LIFE16 ENV/GR/000298). Available online: http://www.biosolids2energy.eu/home.html (accessed on 26 March 2020).

56. Tolkou, A.K.; Zouboulis, A.; Samaras, P. Energy and Wastewater Treatment Plants. In *Sustainability behind Sustainability*; Zorpas, A., Ed.; Nova Science Publisher: Hauppauge, NY, USA, 2014; pp. 1–48.

57. Mamais, D.; Noutsopoulos, C.; Dimopoulou, A.; Stasinakis, A.; Lekkas, T.D. Wastewater treatment process impact on energy savings and greenhouse gas emissions. *Water Sci. Technol.* **2014**, *71*, 303–308. [CrossRef]

58. Massara, T.M.; Malamis, S.; Guisasola, A.; Baeza, J.A.; Noutsopoulos, C.; Katsou, E. A review on nitrous oxide (N2O) emissions during biological nitrogen removal from urban wastewater and sludge reject water. *Sci. Total Environ.* **2017**, *596–597*, 106–123. [CrossRef] [PubMed]

59. Intergovernmental Panel on Climate Change. *Climate Change 2013: The Physical Science Basis*; Working Group I Contribution to the IPCC 5th Assessment Report; IPCC: New York, NY, USA, 2013.

60. Statistics—EUROSTAT. Greenhouse Gas Emissions by Source Sector. Domestic Wastewater; Code: CRF5D1. Available online: https://appsso.eurostat.ec.europa.eu/nui/submitViewTableAction.do (accessed on 26 March 2020).

61. Koutsou, O.P.; Gatidou, G.; Stasinakis, A. Domestic wastewater management in Greece: Greenhouse gas emissions estimation at country scale. *J. Clean. Prod.* **2018**, *188*, 851–859. [CrossRef]

62. Nguyen, T.K.L.; Ngo, H.H.; Guo, W.; Chang, S.W.; Nguyen, D.D.; Nghiem, L.D.; Liu, Y.; Ni, B.; Hai, F.I. Insight into greenhouse gases emissions from the two popular treatment technologies in urban wastewater treatment processes. *Sci. Total Environ.* **2019**, *671*, 1302–1313. [CrossRef]

63. Kornaros, M.; Marazioti, C.; Lyberatos, G. A pilot scale study of a sequencing batch reactor treating urban wastewater operated via the UP-PND process. *Water Sci. Technol.* **2008**, *58*, 435–438. [CrossRef]

64. Malamis, S.; Andreadakis, A.; Mamais, D.; Noutsopoulos, C. Can strict water reuse standards be the drive for the wider implementation of MBR technology? *Desalin. Water Treat.* **2015**, *53*, 3303–3308. [CrossRef]

65. Plevri, A.; Mamais, D.; Noutsopoulos, C.; Makropoulos, C.; Andreadakis, A.; Rippis, K.; Smeti, E.; Lytras, E.; Lioumis, C. Promoting on-site urban wastewater reuse through MBR-RO treatment. *Desalin. Water Treat.* **2017**, *91*, 2–11. [CrossRef]

66. Capodaglio, A.G.; Callegari, A.; Cecconet, D.; Molognoni, D. Small Communities Decentralized Wastewater Treatment: Assessment of Technological Sustainability. In Proceedings of the 13th IWA Specialized Conference on Small Water and Wastewater Systems & 5th IWA Specialized Conference on Resources-Oriented Sanitation, Athens, Greece, 14–17 September 2016.

67. Santana, M.V.; Cornejo, P.K.; Rodríguez-Roda, I.; Buttiglieri, G.; Corominas, L. Holistic life cycle assessment of water reuse in a tourist-based community. *J. Clean. Prod.* **2019**, *233*, 743–752. [CrossRef]

68. Gikas, P.; Ranieri, E.; Sougioultzis, D.; Farazaki, M.; Tchobanoglous, G. Alternative collection systems for decentralized wastewater management: An overview and case study of the vacuum collection system in Eretria town, Greece. *Water Pr. Technol.* **2017**, *12*, 604–618. [CrossRef]

69. Boutsioukis, G.; Fasianos, A.; Petrohilos-Andrianos, Y. The spatial distribution of short-term rental listings in Greece: A regional graphic. *Reg. Stud. Reg. Sci.* **2019**, *6*, 455–459. [CrossRef]
70. Koka, E.; Veshi, D. Irregular Migration by Sea: Interception and Rescue Interventions in Light of International Law and the EU Sea Borders Regulation. *Eur. J. Migr. Law* **2019**, *21*, 26–52. [CrossRef]
71. Tsartas, P.; Kyriakaki, A.; Stavrinoudis, T.; Despotaki, G.; Doumi, M.; Sarantakou, E.; Tsilimpokos, K. Refugees and tourism: A case study from the islands of Chios and Lesvos, Greece. *Curr. Issues Tour.* **2019**, *23*, 1311–1327. [CrossRef]
72. Kaldellis, J.; Kondili, E. The water shortage problem in the Aegean archipelago islands: Cost-Effective desalination prospects. *Desalination* **2007**, *216*, 123–138. [CrossRef]
73. Stathatou, P.M.; Gad, F.K.; Kampragou, E.; Grigoropoulou, H.; Assimacopoulos, D. Treated wastewater reuse potential: Mitigating water scarcity problems in the Aegean islands. *Desalin. Water Treat.* **2014**, *53*, 3272–3282. [CrossRef]
74. EU Leaders Clash over Trillion-Euro Covid-19 Aid in Online Meeting. Available online: https://www.theguardian.com/world/2020/apr/23/clashes-predicted-over-trillion-euro-covid-19-aid-as-eu-meets-online (accessed on 24 April 2020).
75. Wigginton, K.R.; Boehm, A.B. Environmental Engineers and Scientists Have Important Roles to Play in Stemming Outbreaks and Pandemics Caused by Enveloped Viruses. *Environ. Sci. Technol.* **2020**, *54*, 3736–3739. [CrossRef] [PubMed]
76. Save Energy at Home Greece—CA EED. Available online: www.ca-eed.eu (accessed on 24 April 2020).

sustainability

Review

Solar Photocatalysis for Emerging Micro-Pollutants Abatement and Water Disinfection: A Mini-Review

Danae Venieri [1,*], **Dionissios Mantzavinos** [2] and **Vassilios Binas** [3]

1   School of Environmental Engineering, Technical University of Crete, GR-73100 Chania, Greece
2   Department of Chemical Engineering, University of Patras, Caratheodory 1, University Campus, GR-26504 Patras, Greece; mantzavinos@chemeng.upatras.gr
3   Institute of Electronic Structure and Laser (IESL), FORTH, Vasilika Vouton, GR-70013 Heraklion, Greece; binasbill@iesl.forth.gr
*   Correspondence: danae.venieri@enveng.tuc.gr

Received: 26 October 2020; Accepted: 25 November 2020; Published: 1 December 2020

Abstract: This mini-review article discusses the critical factors that are likely to affect the performance of solar photocatalysis for environmental applications and, in particular, for the simultaneous degradation of emerging micro-pollutants and the inactivation of microbial pathogens in aqueous matrices. Special emphasis is placed on the control of specific operating factors like the type and the form of catalysts used throughout those processes, the intriguing role of the water matrix, and the composition of the microbial load of the sample in each case. The interplay among the visible responsive catalyst, the target pollutants/pathogens, including various types of microorganisms and the non-target water matrix species, dictates performance in an unpredictable and case-specific way. Case studies referring to lab and pilot-scale applications are presented to highlight such peculiarities. Moreover, current trends regarding the elimination of antibiotic-resistant bacteria and resistance genes by means of solar photocatalysis are discussed. The antibiotic resistance dispersion into the aquatic environment and how advanced photocatalytic processes can eliminate antibiotic resistance genes in microbial populations are documented, with a view to investigate the prospect of using those purification methods for the control-resistant microbial populations found in the environment. Understanding the interactions of the various water components (both inherent and target species) is key to the successful operation of a treatment process and its scaling up.

Keywords: microorganisms; inactivation; water matrix; catalysts; antibiotic-resistant bacteria; resistance genes

1. Introduction

The current trends in water and wastewater treatment are focused in the development and exploration of environmentally friendly and low-cost technologies. The occurrence of emerging micro-pollutants in the aquatic environment, as well as the presence of various pathogenic microorganisms, impose the application of effective purification methods in order to maintain high hygiene standards and to act toward public health protection.

In this context, advanced oxidation processes (AOPs) have been well studied during the last decades and have proven to be quite promising for the chemical treatment and disinfection of aqueous samples [1–3]. The beneficial action of AOPs is attributed firstly to the in situ generation of highly reactive oxygen species (ROS) like hydroxyl radicals (HO$^\bullet$, E^0 = 1.8–2.7 V), which have the potential to mineralize various organic contaminants contained in waters, classified as bio-recalcitrant [4]. Also, they are capable of causing oxidative stress to "target" microorganisms, exhibiting remarkable biocidal action, as they can lead them to irreversible inactivation [3,5,6]. Encountering the challenge to propose a sustainable technology for the effective treatment of water/wastewater, recent studies have highlighted the application of solar photocatalysis and its variations, which are often used in environmental control processes [7–11]. The prospect of using the solar spectral range and the application at ambient temperature and pressure makes this method even more attractive [12]. Up until now, solar photocatalysis has shown high potential for the degradation of hazardous compounds and the inactivation of multiple microbes present in water and wastewater [1]. Solar and visible light is used for the activation of a substrate, such as a semiconductor photocatalyst, which remains unconsumed after the photoreaction. This activation initiates a chain reaction, through which ROS are produced as pivotal in the degradation of organic compounds and in the inactivation of water pathogens [3,13].

However, the overall performance of the process is highly dependent on various parameters and variables that interfere with the effectiveness, in terms of the satisfactory treatment of water and wastewater. This review paper presents and discusses all possible aspects of solar photocatalysis, according to recent research studies conducted in the field, which deal with this AOP and its lab- or pilot-scale application.

The most important variables that are implicated in the course of the treatment (Figure 1) are (a) the kind of the photocatalytic technique with all recorded operating parameters, (b) the type of microorganism or emerging contaminant, (c) the water matrix, and (d) the special category of antibiotic-resistant bacteria (ARB) or antibiotic resistance genes (ARGs), which have nowadays flooded the interest of the researchers, given their uncontrolled dispersion into the aquatic environment [14,15]. What is commonly accepted is the fact that there is always considerable difficulty in relation to the standardization of the operating parameters, which are applied in each case during such treatments. Apart from the extensive variety of photocatalytic approaches and catalysts that may be used, the main driving forces, which define the final outcome of the treatment, are the wide diversity of organic contaminants and the varied behavior of microbial populations after exposure to the intense conditions of photocatalysis. The latter is more pronounced when ARB and ARGs are included in the frame, considering that both of them may not be fully eliminated post disinfection [16]. In this perspective, the following sections present some of the major issues that are implicated during solar photocatalytic treatment of water and wastewater.

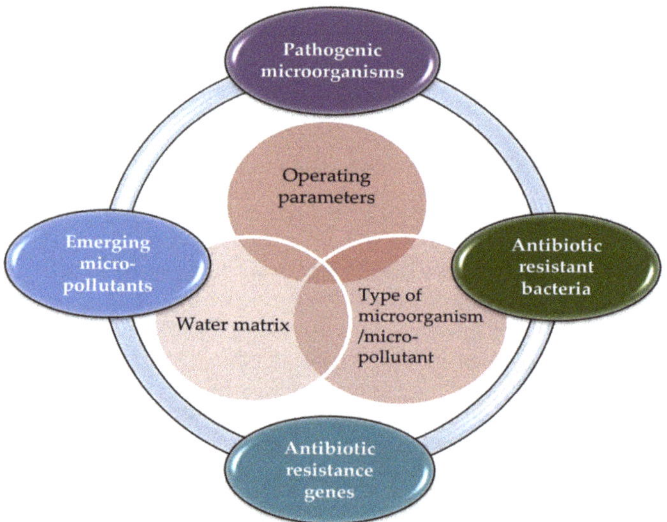

Figure 1. Basic aspects of solar photocatalysis as a treatment method for the degradation of emerging micro-pollutants and the inactivation of waterborne pathogens.

2. Solar Photocatalytic Approaches for Water and Wastewater Treatment

2.1. TiO$_2$ Photocatalysis

Emerging micro-contaminants such as pharmaceuticals and endocrine disruptors are treated ineffectively in conventional wastewater treatment plants (WWTPs), where they are only partially removed through sorption onto the activated sludge, hydrolysis, and biodegradation; because of the low concentration of these micro-contaminants at the ng/L to μg/L levels, though, WWTP operators have not paid particular attention in removing such compounds. Regarding drinking water supplying companies, the use of granular activated carbon alone or combined with ozone, a traditional technique for removing pesticides from waters, can also be effective for other micro-contaminants [17].

The vast majority of solar photocatalytic processes highlight the use of titanium dioxide as an effective catalyst in terms of the degradation and destruction of a wide range of emerging micro-pollutants and microorganisms, respectively [18]. The advantages of heterogeneous semiconductor photocatalysis using TiO$_2$ include its operation at ambient conditions, while among the assets of the catalyst are its low cost, photochemical stability, structural properties, and the fact that it is non-toxic [4]. However, the excitation of this semiconductor requires exposure under irradiation with energy greater than its high band-gap energy (~3.2 eV). This feature makes titania active mainly under the UV spectral range, which is a small fraction of the solar light [19]. Nevertheless, and despite this limitation, there are numerous studies that have investigated the efficiency of pure titania regarding the oxidation of chemical compounds and the inactivation of pathogens under solar light (Tables 1 and 2) [3,20].

Fanourgiakis et al. (2014) studied the simultaneous elimination of synthetic estrogen 17α-ethynylestradiol (EE2) and inactivation of *Escherichia coli* in wastewater, applying simulated solar light and TiO$_2$. According to their findings, the removal rates of EE2 and the bacterium were quite satisfactory, underlining the possible use of pure titania for wastewater purification under solar irradiation [21]. Other attempts that have taken place at the degradation of emerging micro-pollutants have been referred to in investigations that deal with antibiotics. Carbajo et al. (2016), who studied the degradation of multiple antibiotics in water, recorded firstly, the effectiveness of TiO$_2$ upon exposure under solar light, and secondly, the dependence of the catalyst's activity on the concentration of organic pollutants. The total removal of various pharmaceutical compounds occurred in very short periods of time (t$_{30w}$ < 35 min), revealing the beneficial use of titania under specific operational conditions [13]. Similarly, Méndez-Arriaga et al. (2009) used TiO$_2$ for the removal of ibuprofen from water, but they noted an overall enhancement of the process adding H$_2$O$_2$. Nevertheless, the degradation of ibuprofen was significant with TiO$_2$ alone, independently on the solar device employed [22].

Table 1. Selected applications of solar photocatalysis for the elimination of emerging micro-pollutants from aqueous samples.

Emerging Micro-Pollutant		Lab-Scale Photocatalytic Process	Catalyst Concentration/Light Intensity	Aqueous Matrix	Degradation Level	Reference
Endocrine disrupting compounds (EDCs)	Bisphenol-A (BPA) and 17α-ethynylestradiol (EE2)	Various TiO_2 photocatalysts (doping with N, P, Ca, Ag, Na, K, Pt)	125–1000 mg/L/$17.4 \times 10^{-8} - 5.8 \times 10^{-8}$ einstein/(L s)	Wastewater	Up to 90% in 60 min	[2]
	17α-ethynylestradiol (EE2)	TiO_2	500–1000 mg/L/5.8×10^{-7} einstein/(L s)	Wastewater	Up to 89.9% in 90 min	[21]
Pharmaceutical micro-contaminants	Clofibric acid	Silver loaded activated carbon (Ag–AC) nanocomposites	10 mg/L/natural sunlight *	Water	97% after 80 min	[23]
	Diclofenac and memantine	Solar-assisted photocatalysis using hydrothermal TiO_2–SnS_2	5, 27.5 and 50% wt. SnS_2/450 W xenon arc lamp *	Water	59.8% for diclofenac and <5.3% for memantine after 60 min	[24]
	Acetaminophen, ibuprofen and antipyrine	TiO_2-activated carbon heterostructures	250 mg/L/600 W/m^2 (107.14 klx)	Water	Complete conversion within 3–6 h	[25]
	Diclofenac	Immobilized TiO_2-based zeolite composite photocatalyst (TiO_2–FeZ)	Thin films with immobilized composite with TiO_2: FeZ wt% = 74.6: 25.4/124.78 ± 0.11 mW/cm^2	Water	99.7% after 180 min	[26]
	Acetaminophen	TiO_2/activated carbon heterostructures	250 mg/L/600 W/m^2 (107.14 klx)	Water	Complete conversion after 6 h	[27]
	Ibuprofen, acetaminophen and antipyrine	ZnO/sepiolite heterostructured materials	250 mg/L/intensity at 450 W/m^2	Wastewater	70–100% in 10 h	[28]

Emerging Micro-Pollutant		Pilot-Scale Photocatalytic Process (CPC)	Catalyst Concentration/Light Intensity	Aqueous Matrix	Degradation Level	Reference
Endocrine disrupting compounds (EDCs)	Endocrine disruptors	Solar photo-Fenton process assisted with ferrioxalate	Molar ratio Fe/oxalic acid = 3/mean solar intensity = 30 W/m^2	Wastewater	Up to 79% of TOC (total organic carbon) removal in 2 h	[29,30]
Pharmaceutical micro-contaminants	Nalidixic acid	Solar photo-Fenton and biological treatment (immobilized biomass reactor)	20 mg/L of Fe^{2+} and 300 mg/L of H_2O_2/natural sunlight *	Pharmaceutical wastewater	Complete removal after 190 min	[31]
	Atenolol, hydrochlorothiazide, ofloxacin and trimethoprim	Ozone and solar TiO_2-photocatalytic oxidation	Ozone dosage = 18–25 mg/L; TiO_2 P25 = 200 mg L/Q_{UV} up to 40 kJ/L	Water and wastewater	Complete removal of pharmaceuticals and about 70% TOC removal	[32]
	Acetaminophen, antipyrine, caffeine, ketorolac, metoprolol, sulfamethoxazole, carbamazepine, hydrochlorothiazide and diclofenac	Solar heterogeneous photocatalysis with TiO_2, solar photo-Fenton	Ozone concentration in the gas phase = 13 mg/L; TiO_2 = 250 mg/L; Fe(III) = 2.8 mg/L or Fe_3O_4 = 150 mg/L/Q_{UV} = 30–38 kJ/L	Wastewater	80–100% after 180 min	[33]
	Phenol, dichloroacetic acid and pyrimethanil	Two titania (the commercial P25 and a homemade catalyst, TiEt-450)	P25 = 200 mg/L and TiEt-450 = 500 mg/L/mean solar intensity = 30 W/m^2	Deionized water (DW) and natural ground water (NW)	Up to complete removal in t$_{30w}$ = 50–100 min	[13]
	Ofloxacin, sulfamethoxazole, carbamazepine, flumequine and ibuprofen	Two titania (the commercial P25 and a homemade catalyst, TiEt-450)	P25 = 200 mg/L and TiEt-450 = 500 mg/L/mean solar intensity = 30 W/m^2	Deionized water (DW) and natural ground water (NW)	Up to complete removal in t$_{30w}$ = 30 min	[13]
	Ibuprofen	TiO_2	0.1–1 g/L/Q_{UV} up to 60 kJ/L	Water	Total elimination when approximately 80% of TOC still remain in solution (Q_{UV} = 60 kJ/L)	[22]

Table 1. *Cont.*

	Emerging Micro-Pollutant	Pilot-Scale Photocatalytic Process (CPC)	Catalyst Concentration/Light Intensity	Aqueous Matrix	Degradation Level	Reference
	Acetaminophen, antipyrine, atrazine, caffeine, carbamazepine, diclofenac, flumequine, hydroxybiphenyl, ibuprofen, isoproturon, ketorolac, ofloxacin, progesterone, sulfamethoxazole and triclosan	Solar photo-Fenton	Fe = 5 mg/L/mean solar intensity = 30 W/m^2	Wastewater	Complete elimination in t_{30w} = 60–300 min	[34]
	Various contaminants including antibiotics	Conventional photo-Fenton at pH$_3$ and modified photo-Fenton at neutral pH	Fe = 5 mg/L; H$_2$O$_2$ = 50 mg/L/mean solar intensity = 30 W/m^2	Wastewater	Removal of over 95% in t_{30w} up to 150 min	[35]
	Propranolol	TiO$_2$	0.1–0.4 g/L/Q$_{UV}$ up to 1.5 kJ/L	Water	81% in 240 min	[36]
Pesticides	Imazalil, acetamiprid and thiabendazole	TiO$_2$ supported on glass beads	Q$_{UV}$ = 20–40 kJ/L	Wastewater	90–100%	[37]
	Emerging Micro-Pollutant	Pilot-Scale (Solar Raceway Pond Reactors)	Catalyst Concentration/Light Intensity	Aqueous Matrix		Reference
Pharmaceutical micro-contaminants	Contaminants of emerging concern (CECs)—Various antibiotics	Photo-Fenton at neutral pH	Fe^{3+} = 5.6 mg/L and H$_2$O$_2$ = 30 mg/L/mean solar intensity = 30 W/m^2	Wastewater	Over 80% degradation after 15 min	[38]

* Light intensity is not provided.

Table 2. Selected applications of solar photocatalysis for the inactivation of various pathogenic microorganisms in aqueous samples.

	Microorganism	Lab-Scale Photocatalytic Process	Catalyst Concentration/Light Intensity	Aqueous Matrix	Inactivation Level	Reference
Bacteria	*Escherichia coli, Pseudomonas aeruginosa, Bacillus subtilis* and *Staphylococcus aureus*	Silver-loaded activated carbon (Ag–AC) nanocomposites	10 mg/L/natural sunlight *	Water	Satisfactory antimicrobial activity (agar diffusion method)	[23]
	Escherichia coli, Salmonella sp., *Shigella* sp. and *Vibrio cholerae*	Bare and metal-ion (silver, copper and iron)-doped TiO$_2$ photocatalysts	0.1–1.0 g/L/average intensity of radiation = 37.6 mW/cm^2	Wastewater	86.8–100% in 180 min	[39]
	Escherichia coli, total coliforms, Enterococci, *Vibrio owensii, Vibrio alfacsensis* and *Vibrio harveyi*	Ink-jet printed composite TiO$_2$/SiO$_2$ thin film	UV dose up to 44.91 Wh/m^2	Water/marine water	Up to 99% in 100 min	[40]
	Escherichia coli, Pseudomonas aeruginosa and *Bacillus cereus*	N-doped TiO$_2$ photocatalysts	25–100 mg/L/irradiance = 1.31 × 10^{-2} W/m^2	Water	4–6 Logs in 15–60 min (depending on the bacterium and catalyst concentration)	[9]
	Escherichia coli and *Klebsiella pneumoniae*	Mn-, Co- and Mn/Co-doped TiO$_2$ catalysts	25–250 mg/L/irradiance = 1.31 × 10^{-2} W/m^2	Water	5–6 Logs in 15–30 min (depending on the bacterium and catalyst concentration)	[19]
	Staphylococcus aureus	Fe-, Al- and Cr-doped TiO$_2$ catalysts	5–50 mg/L/irradiance = 1.31 × 10^{-2} W/m^2	Water	5 Logs in 6–30 min	[41]
	Vibrio cholerae	Ag@ZnO core–shell-structured nanocomposites	0.5 mg/L/ *	Water	Almost 98% in at 40–60 min	[5]
	Escherichia coli	Ag core–TiO$_2$ shell-structured (Ag@TiO$_2$) nanoparticles	0.4 g/L/average light intensity = 970 × 10^2 lux	Water	8 Logs in 15 min	[42]
	Escherichia coli	TiO$_2$ P-25, PC500, Ruana and Bi$_2$WO$_6$	0.05–1 g/L/solar UV irradiance up to 40 W/m^2	Water	6 Logs in 15–120 min	[6]

Table 2. *Cont.*

	Microorganism	Lab-Scale Photocatalytic Process	Catalyst Concentration/Light Intensity	Aqueous Matrix	Inactivation Level	Reference
	Heterotrophic bacteria	TiO_2	0.5 g/L/solar UV irradiance = 14–27 W/m^2	Dairy wastewater	41–97% in 30 min	[43]
	Escherichia coli	Neutral solar heterogeneous photo-Fenton (HPF) over hybrid iron/montmorillonite/alginate beads	0–20 beads with 10 ppm of H_2O_2/irradiation intensity = 1200 W m^{-2}	Water	7 Logs in 1 h	[44]
Fungi	*Fusarium* sp.	TiO_2	35 mg/L/average solar UV-A irradiance = 25.78 W/m^2	Water	3 Logs in 1-6 h	[45]
Viruses	Phages MS2, UX174 and PR772	TiO_2	50 mg/L/average solar UV-A irradiance = 19–33 W/m^2	Water	3 Logs in 1 h	[46]
	Phage MS2	Mn-, Co- and Mn/Co-doped TiO_2 catalysts	100 mg/L/average solar UV-A irradiance = 12.7–13.4 W/m^2	Wastewater	Up to 4 Logs in 60 min	[47]
	Phage MS2	Iron (hydr)oxide-mediated Fenton-like processes	200 mg/L of iron oxides and 50 μM of H_2O_2/irradiance = 320 W/m^2	Water	5 Logs in 240 min	[48]
Parasites	*Cryptosporidium parvum* oocysts	TiO_2	100–200 mg/L/irradiance = 500 W/m^2	Water and simulated WWTP effluent	99 up to 50% in 5 h, in water and wastewater, respectively	[10]

	Microorganism	Pilot-Scale Photocatalytic Process (CPC)	Catalyst Concentration/Light Intensity	Aqueous Matrix	Inactivation Level	Reference
	Escherichia coli, Enterococcus faecalis	Synthetized Ag modified $BiVO_4$ composite	0.2–1 g/L/average solar UVA irradiance = 27 ± 2 W/m^2	Water and secondary effluents	6 Logs in 60 min	[12]
	Enterococcus faecium and *Klebsiella pneumoniae*	Immobilized TiO_2 reduced graphene oxide	loading ofTiO_2-rGO = 0.89 mg/cm^2/average solar UVA irradiance up to 40 W/m^2	Rainwater	8–9 Logs in 240 min	[49]
	Escherichia coli K12	Industrial TiO_2-coated paper matrix fixed on a tubular support in the focus of the CPC	Coated at a dose of 20 g TiO_2/m^2/average solar UVA irradiance = 22 W/m^2	Water	6 Logs in 90 min	[50]
Bacteria	*Escherichia coli*	TiO_2 (Degussa P25) in suspension or TiO_2 supported on Ahlstrom paper (NW10) fixed	50 mg/L (suspension) and dose = 11.8 g/m^2 (fixed)/average solar UVA irradiance up to 40 W/m^2	Water	Up to 6 Logs in 50 min	[51]
	Escherichia coli	Suspended TiO_2	100 mg/L *	Urban and simulated urban effluents	Up to 5.5 Logs when Quv = 12 kJ/L	[52]
	Aeromonas hydrophila	Thin film fixed-bed reactor (TFFBR) coated with P25 DEGUSSA TiO_2	Density of TiO_2 = 20.50 g/m^2/solar irradiance = 980–1100 W/m^2	Water	Up to 5 Logs in 30 min, depending on operating conditions	[20]
Fungi	*Fusarium solani*	Synthetized Ag modified $BiVO_4$ composite	0.2–1 g/L/average solar UVA irradiance = 27 ± 2 W/m^2	Water and secondary effluents	Almost 3 Logs in 240 min	[12]
	Fusarium solani	Suspended TiO_2	100 mg/L *	Urban and simulated urban effluents	Up to 2.5 Logs when Quv = 5 kJ/L	[52]

	Microorganism	Pilot-Scale (Solar Raceway Pond Reactors)	Catalyst Concentration/Light Intensity	Aqueous Matrix	Inactivation Level	Reference
Bacteria	Total Coliforms, *Escherichia coli* and *Enterococcus* sp.	Solar photo-Fenton	Fe^{2+} = 2.5–20 mg/L and H_2O_2 = 30 or 50 mg/L/average solar UVA = 13–34 W/m^2	Secondary wastewater effluents	4–5 Logs in 11–15 h	[53]

* Light intensity is not provided.

The biocidal effect of UVA irradiation has been long documented and is attributed to its absorption by cellular components called chromophores, causing further damage and oxidative stress to the microorganisms [4,54]. During photocatalysis the progressive generation of ROS results in detrimental effects on microbial components and on the cellular layers, beginning from the cell wall and the outer membrane. Afterwards, the lesions expand toward the inner proteins (enzymes) and the nucleic acids (genetic material) [43,55]. The primary species responsible for microbial destruction is hydroxyl radical ($HO^\bullet$) followed by superoxide radical anion ($O_2^{\bullet-}$), hydro-peroxyl radical ($HO_2^\bullet$), and hydrogen peroxide (H_2O_2) [56]. Selected applications of TiO_2 for solar disinfection purposes may be seen in Table 2. The bactericidal action of titania proved to be significant when *E. coli* was the target organism, either in water or in wastewater [6,52]. Catalyst concentrations up to 0.5 g/L seem to be sufficient for the complete removal of the bacterium with an initial concentration of 10^6 CFU/mL. Parallel performance has been observed regarding other microorganisms as well, like fungi (*Fusarium* species) [45,52] and heterotrophic bacteria in dairy wastewater [43].

During recent years, though, the trend in the broad solar photocatalytic area has been to develop and explore newly synthesized materials that potentially could serve as efficient catalysts for the process. The general concept is to improve the activity of titania and to expand its absorption spectrum toward the visible light region. Screening the current literature, various materials have arisen that show promising performance in terms of the elimination of emerging micro-pollutants and waterborne pathogens (Tables 1 and 2). Many different strategies have been adopted for either morphological or chemical modifications of the catalyst [57,58]. Those include modifications of TiO_2 surface with noble metals or other semiconductors or incorporation of additional components in the catalyst structure, like non-metal or/and noble and transition metal deposition [57]. The performance of modified titania is highly improved under simulated and natural solar light, and better removal rates are achieved with various contaminants/pathogens.

In this perspective several attempts have been made using doped-titania materials, in terms of the degradation of hazardous and emerging micro-contaminants in water and wastewater. For example Dimitroula et al. (2012) proceeded with the removal of bisphenol-A (BPA) and 17α-ethynylestradiol (EE2) from wastewater, using various TiO_2 photocatalysts doped with N, P, Ca, Ag, Na, K, and Pt. The overall photoactivity of modified titania under visible light was enhanced, but the treatment performance was not improved substantially [2]. This outcome verified the fact that modified titania do not always work well under the operating conditions in each case. Besides, there have been reports of many possible limitations to metal-doped titania materials, like photo-induced corrosion and promoted charge recombination at some metal sites [57,59]. On the other hand, more applications have been overviewed in recent studies, regarding the use of doped-titania and disinfection processes. For instance, various metal-doped TiO_2, such as Fe–, Mn–, Co– or Al–TiO_2 catalysts, have been used successfully for the inactivation of bacteria (*E. coli*, *Klebsiella pneumoniae*, *Staphylococcus aureus*) and viruses (bacteriophage MS2) in water and wastewater [19,41,47]. In all those cases microbial inactivation was 2–3 times faster, compared with the respective occurring with the pristine P25 TiO_2. The improved activity of metal-doped titania was credited to the optical absorption shifts toward the visible region and to the recombination delay of the electron–hole pair. Also, Sreeja et al. (2017) investigated the performance of Ag core–TiO_2 shell-structured (Ag@TiO_2) nanoparticles and found that those catalysts were quite efficient for the inactivation of *E. coli* in water under solar light irradiation [42]. Complete disinfection (8 Log reduction of the bacterial population) was achieved within 15 min of treatment applying 0.4 g/L Ag@TiO_2 catalyst loading. Interestingly, similar promising results were obtained testing other species such as *Bacillus cereus* with N-doped TiO_2 photocatalysts using various nitrogen precursors (urea, triethylamine-TEA, and NH_3) [9]. Although *B. cereus* exhibited high resistance, N-doped TiO_2 catalysts were more active than pure titania in water samples and under simulated solar irradiation. Generally, the use of modified titania, at least in the case of water/wastewater disinfection under solar irradiation, seems to be faster than other treatment techniques, highlighting the competitive nature of the proposed process against more conventional disinfection systems.

2.2. Slurry or Immobilized Catalysts?

One of the major concerns or debates refers to the choice of the catalyst form to be used for environmental applications. Catalysts in slurry phase are well-known for their effective performance and rather popular; however, such processes require further treatment steps so as to remove the catalyst from the treated sample (water or effluent). On the other hand, another option is to immobilize the catalyst onto appropriate surfaces, surpassing the need for post-treatment handling [37]. Nevertheless, even then, other issues are implicated in the oxidation process, like the decrease of the surface area of the catalysts that is available for the photocatalytic reactions [60]. This feature results in lower degradation rates of chemical compounds and slower inactivation of microbial pathogens in aqueous matrices when immobilized catalysts are employed, compared with the suspended systems [61]. The choice of the catalyst form should be weighed carefully, based on the treatment that is to be applied and on the special requirements in each case (type of pollutant/microorganism, initial concentration water matrix, etc.).

Salaeh et al. (2016) investigated the possibility of removing diclofenac from water using immobilized TiO_2-based zeolite composite photocatalyst (TiO_2–FeZ) and simulated solar light. Diclofenac was removed by 80.1% after 15 min of exposure, with the adsorption of the pharmaceutical playing the most significant role in the overall treatment efficiency [26]. Also, in another study TiO_2 supported on glass beads was tested for tertiary treatment of residual pesticides, achieving rates over 90%, but only with the additional contribution of hydrogen peroxide as an electron acceptor [37]. Respective attempts have been made in the field of disinfection. Khan et al. (2012) worked with a thin-film fixed-bed reactor (TFFBR) for the inactivation of aquaculture pathogen *Aeromonas hydrophila*, demonstrating that high sunlight intensities (>600 W/m^2) and low flow rates (4.8 L/h) play key role in the inactivation of this fish pathogen [20]. Sichel et al. (2007) achieved a 6 Log reduction of *E. coli* within 90 min, using TiO_2 immobilized on Ahlstrom paper in a compound parabolic collector (CPC reactor), highlighting that low flow rates contribute to a more efficient photocatalytic disinfection [50].

In an attempt to improve the photocatalytic activity when TiO_2 films are used and to counterbalance any loss that may occur, many researchers propose the application of an external electric bias. Dunlop et al. (2008), who worked with spores of *Clostridium perfringens* and TiO_2/Ti films (working electrode), proved that applying an external bias of 1 V led to 60–70% higher inactivation rates, while when no bias was applied the disinfection efficiency was inadequate [62]. Based on their research, the potential gradient forces the electrons toward the cathode, thus minimizing the rate of electron–hole recombination.

2.3. Photocatalysts Other than TiO_2

Titania nanoparticles and its composites show remarkable results during solar photocatalysis of water and wastewater. Especially the metal and non-metal-doped nanoparticles have been extensively used for multiple applications, demonstrating promising prospects of a "clean and green" aquatic environment. Nevertheless, we should not overlook some other semiconductors that have emerged as alternative approaches in this field of treatment and disinfection.

Zinc oxide nanoparticles with a wide band-gap of 3.37 eV appear to be a nice option, considering some recorded assets, such as good optoelectronic, piezoelectric, and catalytic properties [28]. However, photo-corrosion may worsen the performance of ZnO, causing limited stability. Therefore, some researchers have tested the use of supplementary materials as support to ZnO nanoparticles. For example, ZnO-supported clays have been prepared for photocatalytic applications, like ZnO/sepiolite heterostructured materials. Akkari et al. (2018) used those composited for solar photocatalytic degradation of pharmaceuticals in wastewater. According to their findings, ibuprofen, acetaminophen, and antipyrine were readily degraded in wastewater, indicating the superiority of those materials compared to other catalysts used for solar photocatalysis [28]. ZnO nanocomposites have also been used successfully for disinfection purposes of various bacterial species like *E. coli*, *Vibrio cholerae*, and multi-drug-resistant *Bacillus* sp. [5,63,64]. Given that the solar photocatalytic activity of the metal oxide nanostructures is increased by formation of metal/metal oxide hybrid structures, Das et al. (2015) synthesized Ag@ZnO core–shell structure nanocomposites and tested their potential to inactivate *V. cholerae* in water. The results showed that this highly pathogenic bacterium may decrease up to 98% after 40–60 min of sunlight exposure with a catalyst loading of 0.5 mg/L [5]. The same group worked with Ag@SnO$_2$@ZnO core–shell nanocomposites and Fe-doped ZnO nanoparticles, as well, studying their biocidal properties against *Bacillus* sp. and *E. coli*, respectively. In both cases the synthesized materials exhibited satisfactory performance in terms of the inactivation of pathogens in water (Tables 2 and 3) [63,64]. In all those cases catalysts had a stable structure and no silver leaching was observed.

Further attempts have been made to explore more catalysts with acceptable solar performance. In this sense, cadmium sulfide (CdS) seems to be quite effective regarding the disinfection of aqueous matrices with high concentrations of *E. coli* and *S. aureus* under visible light [65]. Silver orthophosphate (Ag$_3$PO$_4$) is a low band-gap photocatalyst with enormous potential in harvesting solar energy. What is important regarding this catalyst is that it is characterized by a low electron–hole recombination rate, but with low long-term stability, as it is decomposed in the absence of sacrificial agent [66]. In this case, leaching of silver in the liquid phase may contribute to disinfection through homogeneous reactions. This drawback may be surpassed by synthesizing various Ag$_3$PO$_4$-based composites. Ag$_3$PO$_4$ and Ag$_3$PO$_4$/TiO$_2$ materials have the potential to achieve good inactivation rates of *E. coli* under solar irradiation, while other studies present the disinfection efficiency of several Ag$_3$PO$_4$/TiO$_2$ composites against multiple pathogens [67–69].

Among the numerous visible light active photocatalysts bismuth vanadate (BiVO$_4$) has received attention despite the fact that very few water disinfection studies have been reported. Its activity alone is not that significant, as the recombination rate of photo-induced electron–hole pair is really fast and high. Metal deposition on the surface of the catalyst seems to work toward overcoming this drawback, leading to enhanced activity under solar light. In this perspective, the silver deposition on the surface of BiVO$_4$ made this catalyst capable of inactivating three waterborne pathogens, namely, *E. coli* (Gram-negative bacteria), *Enterococcus faecalis* (Gram-positive bacteria) and spores of *Fusarium solani* (phytopathogen) under natural sunlight [12]. Finally, one more catalyst reported in current literature is Bi$_2$WO$_6$, which has the advantage of absorbing more solar photons. This catalyst has the potential to accelerate the bactericidal action of solar irradiation, given that a concentration of 0.5 g/L is sufficient for a 6 Log reduction of *E. coli* in water within 105 min [6].

2.4. Heterogeneous Photo-Fenton Systems

Among the AOPs applied for water and wastewater treatment, the photo-Fenton process has become very popular as an eco-friendly choice for organics mineralization and microbial inactivation. This process takes place in the presence of ferrous or ferric salts and hydrogen peroxide in acidic media, and hydroxyl radicals are generated through the Fe^{2+}/Fe^{3+} redox cycle. The production of hydroxyl radicals is greatly enhanced under UV–vis irradiation, as transformation of Fe^{3+} to Fe^{2+} is promoted. The main challenge when applying this method is to operate at neutral or near-neutral conditions and not in the range of 2.5–3.5, which is optimum for this AOP [70]. This pH range is prohibited for environmental applications, and further actions should be taken post treatment and prior to the disposal of treated streams into the aquatic bodies (e.g., neutralization).

In this view, current research studies have proposed heterogeneous Fenton-like systems, which operate well and efficiently at neutral or near-neutral conditions (Table 2). New organic or inorganic supports have been tested for the catalysts used in photo-Fenton processes, especially biopolymers like sodium alginate, which is biocompatible, inexpensive, and can be easily assembled into spherules or beads. Barreca et al. (2015) synthesized iron-enriched montmorillonite alginate beads for the inactivation of *E. coli* and recorded a 7 Log reduction at pH 7 after 60 min under solar irradiation with 10 mg/L H$_2$O$_2$ [44]. Also, other materials served as efficient catalysts for the removal of MS2 coliphage from water at neutral conditions [48]. This phage was inactivated successfully in water in the presence of hematite (α-Fe$_2$O$_3$), goethite (α-FeOOH), and magnetite (Fe$_3$O$_4$) and under solar light, and all materials exhibited stability with negligible iron leaching. Also, promising results have been derived regarding the wastewater treatment by means of the photo-Fenton process. De la Obra Jiménez et al. (2019), who worked with raceway pond reactors, observed total inactivation of total coliforms *E. coli* and *Enterococcus* sp. in wastewater secondary effluents in continuous flow and neutral pH within 60 min in the presence of 50 mg/L H$_2$O$_2$ [53].

Similar studies may be overviewed regarding the degradation of emerging micro-pollutants (Table 1). Solar photo-Fenton reactions are capable of removing endocrine disruptors (EDCs) and various antibiotics from water and wastewater, either alone or combined with other processes [29,31,33,38]. For instance, Sirtori et al. (2009) investigated the degradation rate of nalidixic acid, which belongs to the quinolone group of antibiotics, by means of photo-Fenton and biological treatment. Photo-Fenton was found to be a successful enhancer of the biodegradability of wastewater, acting as a supplementary technique to an immobilized biomass reactor in order to achieve mineralization and detoxification of industrial wastewater [31]. Moreover, Soriano-Molinao et al. (2019) accomplished the removal of 80% of the concentration of chemicals of emerging concern from wastewater after 15 min of photo-Fenton at circumneutral pH in solar raceway pond reactors [38]. Based on all those results, heterogeneous photo-Fenton systems at neutral pH seem to be a feasible solution for water/wastewater treatment with acceptable results, without causing any disturbance or toxicity to the surrounding environment.

2.5. Transformation By-Products

Solar photocatalysis of contaminants may result in the formation of transformation by-products (TBPs) that are less biodegradable and/or more toxic than the original compound. This is more likely to happen if the experiments have been performed in environmental matrices rather than pure water (as this is mainly the case for the studies shown in Table 1), since less biogenic TBPs may also be generated from photocatalytic transformations involving the non-target species inherently present in the matrix (i.e., the effluent organic matter typically found in treated wastewaters and the natural organic matter found in groundwaters) [31]. The level of toxicity induced by the generation of by-products is often unpredictable and sometimes related to the duration of the process. The toxicity in short treatments usually decreases gradually in the course of the photodegradation [4].

The effect of photocatalysis on the properties of the effluent is usually assessed by means of biodegradability and/or toxicity tests. The standard BOD (biochemical oxygen demand) test is commonly employed as a measure of aerobic biodegradability, which is also assessed by means of shake flask tests, respirometry, and the Zahn–Wellens test [24]. Anaerobic biodegradability tests are less popular and usually measure the rate of biogas production. Acute toxicity is usually assessed against freshwater and marine microorganisms and the results are usually quoted in the form of EC50 values [26]. It should be pointed out that identification of TBPs, although conceptually advantageous, may not be feasible even when sophisticated analytical tools are available. This is due to the fact that the concentration of micro-contaminants may be 2–3 orders of magnitude lower than the organic and inorganic, non-target matrix components and, therefore, interferences mask the presence of TBPs in the matrix [31].

3. The Intriguing Role of the Water Matrix

The water matrix that is mainly used in research studies dealing with AOPs and water/wastewater treatment is ultrapure water. This choice is based primarily on the need to gain fundamental understanding of processes such as degradation kinetics, mechanisms, and pathways without taking into account the impact and the interference of the water matrix effect. Notwithstanding, the latter may be extremely influential to the overall performance of each technique and has the potential to lead to an unreliable outcome.

It is well established that a high level of the water matrix complexity causes deterioration of AOPs' efficacy. This occurs because the pollutants/microorganisms and the ingredients of the matrix (e.g., dissolved organic matter, inorganic constituents, etc.) develop a competitive action toward the generated ROS or the active sites of the catalysts/activators when heterogeneous processes are applied [71]. In this sense, for example, in a case of sulfamethoxazole degradation using solar photocatalysis over WO_3/TiO_2 suspensions, the pseudo-first order kinetic constant decreases as the matrix shifts from ultrapure water to drinking water (DW: containing bicarbonates and other ions) and finally to secondary treated wastewater (WW: containing residual organics and various ions) [72]. On the other hand, the exact reverse behavior may take place under different operating conditions and when other contaminants are to be degraded, like bisphenol-A (BPA); the highest rates of BPA degradation are recorded when the sample is wastewater, compared with other matrices, like ultrapure water [73].

Apparently, the target micro-pollutants/microorganisms that are to be degraded/inactivated, the constituents of the matrix, the ROS, and the catalysts/activators, if they are present, develop tricky and challenging interactions among them with unpredictable results. Eventually, the nature of those interactions will define any reaction kinetics and mechanisms through a synergy or an antagonism, which may be generated. Moreover, the relative contribution of each individual effect may depend on the specific treatment system in question and, for a certain system, on the specific operating conditions.

Nevertheless, some cases underline the fact that the effect of water matrix on photocatalytic disinfection/degradation is case specific. The mechanisms and kinetics of photocatalytic disinfection are highly affected by the presence of inorganic ions (e.g., bicarbonates, chlorides etc.), organics (e.g., natural organic matter (NOM)) and suspended solids. Those components aid in the resistance of microorganisms, considering that they act as physical shields that interfere in the whole process [74]. That is why wastewater has always been dealt with as an aqueous matrix of special attention with special complexity and intrinsic features. Zuo et al. (2015) presented the deterioration of photocatalytic disinfection of *E. coli* due to the presence of ammonia and nitrites in the matrix. The overall effect was attributed to the partial consumption of hydroxyl radicals during the conversion of inorganic nitrogen to nitrates [75]. Similar observations were made by Marugán et al. (2010), who recorded the unfavorable effect of carbonates, phosphates, and humic acid on the inactivation of *E. coli* [76]. However, they highlighted the positive effect of chlorides on disinfection, which may further contribute to the production of toxic organochlorinated by-products. The latter may counterbalance the loss of hydroxyl radicals, leading to an improvement of disinfection efficiency. What was even more surprising was that the same components seemed to slow down the photocatalytic degradation of dyes, making the whole issue of "the water matrix effect" rather a "brain teaser" with an unpredicted outcome. The main suggestion in the literature is the careful standardization of operating conditions in each case, based on the special features of the chemical pollutants and microbial pathogens contained in the sample.

4. Type of Waterborne Pathogens Tested in Solar Photocatalysis

Water and wastewater contain a remarkably extensive variety of microorganisms, belonging to different groups with diverse structures and features. The latter affect inevitably the microbial response and their overall behavior during a disinfection process, as well as the specific mode of their inactivation. According to the recent literature many studies have been conducted so as to provide insight about the principles and mechanisms of microbial inactivation. However, there is still a lot to be revealed and clarified. Screening indicative published data, it is quite obvious that most of the disinfection studies related to solar photocatalysis are focused on the investigation of bacterial species and spores (Table 2), leaving out other virulent pathogens, which are important to public health. What is more is that although multiple bacterial species are contained in water and wastewater, the one that is always mentioned in disinfection applications is the well-known *E. coli* [6,42,44,51]. Nevertheless, focusing on just one bacterial indicator poses the risk of extracting biased conclusions in terms of the effectiveness of solar photocatalytic applications.

The extent up to which cell (or other) damages occur varies greatly, depending on the type of microorganism tested each time. Therefore, in the case of bacteria the level of damages and cell permeability caused by ROS are defined, among other parameters, by the thickness of the cell wall. The main differences are identified between Gram-positive and Gram-negative species, as the first ones possess a thick cell wall that contains many layers of peptidoglycan and teichoic acids. Those components provide the potential of preserving their viability during photocatalytic treatment, as the penetration of free radicals is rather obstructed [40]. However, the higher resistance of Gram-positive bacteria is not always confirmed, as the operational conditions and the bacterial indicators employed in each case may reverse this precedence order [9]. In this sense, there are cases where high catalyst concentrations may be required up to 300 mg/L for the complete inactivation of Gram-negative bacteria [77]. The role of cell wall structure and complexity in the overall behavior of bacteria during photocatalysis is still under investigation and many parameters are yet to be explored. It is commonly accepted though, that the disinfection efficiency of a process should be assessed using representative indicators of both groups of bacteria, in order to obtain reliable and accurate results and an objective overview of the process' limits.

Another issue under consideration is the cellular form of the target microorganism. For instance, some pathogenic bacteria are found in the aquatic environment in the form of endospores, which are considered really resistant under the stressed conditions of disinfection. Endospores contain a thick coating made by proteins, which usually require prolonged treatment and exposure under solar irradiation. García-Fernández et al. (2015) studied the effect on the microorganism type of the solar photocatalytic treatment and found that vegetative cells are much more sensitive than spores. In that specific case *Fusarium* spores (fungus) were tested, which showed remarkable resistance to TiO_2 photocatalysis due to rigid structures composed of polymeric sugars, proteins, and glycoproteins. Also, their wall contains an outer xylan layer that confers significant resistance to oxidative stress [52]. In another case, *Clostridium perfringens* spores with a dipicolinic acid–calcium–peptidoglycan complex could be harmed only by hydrogen peroxide, which can be further activated by ferrous ion that is incorporated into the spore coating. This process is called in vivo Fenton reaction [62].

The waterborne protozoa constitute another group of pathogenic microorganisms that are found in the aquatic bodies in the resistant form of cysts/oocysts. *Cryptosporidium parvum* and *Giardia lamblia* are considered very virulent with extremely low infectious dose and yet they have not been mentioned frequently in the literature in relation to disinfection techniques. Generally, both protozoan species show significant tolerance during conventional methods, like chlorination, but also during many AOPs [74]. Oocysts of *C. parvum* require up to 5 h for a substantial decay and removal from distilled water during TiO_2 solar photocatalysis [10]. Moreover, the authors stated that because of the robustness of the oocysts, *C. parvum*'s inactivation would probably ensure the elimination of other less resistant pathogens. Even if oocysts remain as residual microorganisms after treatment, they are not considered infective as excystation occurs with the subsequent generation of sporozoites. The combination of solar light with a catalyst causes destruction of the oocyst cell walls, and the final picture is empty cells characterized as "ghosts", which remain after the process [70].

Much less research has been conducted on the photocatalytic inactivation of viruses, whose significant presence in the aquatic environment verifies their resistant nature and tolerance during conventional disinfection methods. Up until now, studies demonstrated the existence of such viruses in treated effluents, highlighting the inadequacy of conventional purification methods [78]. Viruses are traditionally known to maintain their structural properties and infectivity when hostile conditions are induced in the surrounding area [79]. Upon application of a photocatalytic process, viral inactivation may occur only when substantial oxidizing power is provided, which is necessary for the deformation of their protein capsid and the development of lesions in their nucleic acid. The absence of any enzymes or other typical cellular structure leaves capsid and genetic material as the only targets of the ROS generated during AOPs techniques [47]. Viral adsorption and general adherence onto the catalysts' nanoparticles is the first step of their inactivation in photocatalytic processes, followed by the attack on the protein capsid and other binding sites of the viruses [80]. On the other hand, certain studies proposed a different mode of action and mechanism of photocatalysis against viruses. What mostly occurs is the interaction between free hydroxyl radicals in the bulk phase and the viruses, as electrostatic repulsion does not allow the interaction and close contact between the catalyst and the virus [46]. The application of a positive potential to an immobilized TiO_2 electrode may induce an electrostatic attraction between the catalyst and the viral capsid, which is mostly negatively charged. Also, Fenton's reagent and metal-doped titania seem to eliminate successfully MS2 coliphages, as reduction up to 5 Logs may occur within 60 min of treatment [47].

The final target of ROS in the course of photocatalysis is the genetic material of microorganisms and viruses (DNA or RNA). Nucleic acids are rather susceptible to the produced oxidative power through attacks either at the sugar or at the base [81]. All damages and lesions in the microbial genetic material are subject to restoration in the case of some bacterial species, according to their properties. This feature, the so-called "photoreactivation," is the main disadvantage of photocatalytic treatment and generally of the processes that utilize UV irradiation. Some bacteria have the potential to repair any destruction sites or "mistakes" on their genetic material through special enzymatic activity. Such enzymes mainly act under light (300–500 nm) and split the dimmers formed as a consequence of irradiation [82]. Although restoration activity takes place usually after exposure to UV-C irradiation, it has also been reported when UV-A is employed, involving not only bacteria but other microbes like protozoan cysts [83]. Therefore, having in mind that solar photocatalytic processes have no residual action, it is crucial to design properly such applications in order to ensure the durability of the disinfection and the inability of waterborne pathogens to proliferate post treatment. Catalyst loading, light energy, time of irradiation are some of the parameters that must be defined and standardized properly to cause irreversible damages in microbial components and structures. The possibility of microbial reactivation always remains, but it should be minimized, though, for public health protection and if solar photocatalysis is to be applied for water disinfection purposes.

5. Antibiotic-Resistant Bacteria (ARB) and Antibiotic Resistance Genes (ARGs)

A special group of microorganisms contained in water and wastewater is referred to the antibiotic-resistant bacteria (ARB), which have already attracted much scientific attention nowadays. The effective application of a disinfection process should always include this specific target microbial group, as it raises many concerns about human health. The uncontrolled use of antibiotics in medical, veterinary, or even agricultural practices and their incomplete removal in WWTPs has led to their uncontrolled excretion in the environment, resulting in an excessive rise of antibiotic resistance in various bacteria by the dissemination of antibiotic resistance genes (ARGs) [84]. ARB and ARGs seem to prevail in the aquatic environment, inducing further resistance within microbial communities, while they have also been documented as emerging contaminants. Many different kinds of ARGs have already been detected in aquatic systems, including WWTPs, and their effluents verify their persistence during treatment (Table 3). Water bodies and particularly WWTPs are extraordinary settlements for the proliferation of ARB and the dissemination of ARGs through horizontal transference of genetic elements, conferring resistance to multiple antibiotic compounds [16]. The main concern and question is whether current treatment processes and disinfection approaches are capable of removing all ARB and ARGs present in water/wastewater, prohibiting their revival in effluents. According to the current literature this is quite common and many multi-drug-resistant bacteria, as well as ARGs, have been detected in the end-streams of WWTPs [85,86]. Moreover, in some cases ARGs are increased in the course of treatment, resulting in extremely high concentrations in the effluents [87].

Therefore, what is mostly needed is the establishment and application of effective technologies toward the control of ARB and the elimination of ARGs from water/wastewater. Failure to limit their dispersion into the aquatic environment threatens public health and contributes to a further increase of resistant populations. The extent to which treatment and disinfection processes inactivate ARB and eliminate the genes relevant to resistance is still under discussion [88]. The question that arises is how far and under which operational conditions does disinfection eliminate ARB and ARGs.

While the risk still exists, solar photocatalysis seems to work well in this direction, providing promising results regarding the inactivation of ARB (Table 3). As already mentioned, this method overcomes many disadvantages of conventional purification processes like the toxic by-products of chlorination or certain action limitations of UV irradiation, which have the potential to remove ARB from water and wastewater [89,90]. Doped-titania materials have the potential to inactivate sufficiently antibiotic-resistant *E. coli* or *K. pneumoniae* [15,91]. Metal and non-metal-doped TiO_2 under solar irradiation led to up to 6 Log bacterial reductions within 60 min of treatment of urban wastewater. Also, Venieri et al. (2016) studied the possible changes in the antibiotic resistance profile of *K. pneumoniae* post treatment and found out that in some cases residual cells after disinfection were more susceptible in specific antibiotic compounds [15]. The same authors documented the simultaneous loss of *K. pneumoniae*'s ARGs in the course of photocatalysis. Fe-doped ZnO nanoparticles and Ag@SnO_2@ZnO core–shell nanocomposites exhibited similar performance, adequately inactivating *E. coli* and *Bacillus* sp. in water, respectively [63,64]. Neither bacteria regrew after treatment and *Bacillus* sp. lost substantial resistance. Also, comparing the effectiveness of Ag@SnO_2@ZnO core–shell nanocomposites with traditional chemical disinfectants and UV-250 nm, it was found that they had lesser impact on the resistance profile of the bacteria.

The elimination of ARGs during solar photocatalysis has been underreported in recent'studies. Although there are data regarding their prevalence in water and wastewater (Table 3), more information is needed about their response in the presence of a semiconductor and solar light. Furthermore, given that ARGs are mostly carried in bacterial plasmids, special attention should be paid to the persistence of plasmids and their integrity level during treatment. According to Mao et al. (2015), the optimum removal of ARGs from wastewater requires high irradiation intensities or the combination of UV with a photocatalytic treatment [87]. Up until now, the point of agreement is that wastewater is an important repository of ARGs that needs more effective treatment than conventional applications.

Table 3. Elimination of antibiotic-resistant bacteria (ARB) and antibiotic resistance genes (ARGs) present in the aquatic environment by means of solar photocatalysis and other disinfection methods (indicative recent literature).

Antibiotic Resistant Bacteria (ARB)	Aqueous Matrix	Treatment	Removal Level	Reference
Escherichia coli	Wastewater	Solar TiO_2 photocatalysis	93.17% removal after 10 min	[11]
Escherichia coli	Water	Solar photocatalysis using Fe-doped ZnO nanoparticles	More than 99.9% removal after 90 min	[63]
Escherichia coli	Wastewater	Solar photocatalysis using N-doped TiO_2 nanoparticles	More than 5 Log bacterial reduction after 10 min of irradiation	[91]
Klebsiella pneumoniae	Wastewater	Solar photocatalysis using Mn-, Co- and binary Mn/Co-TiO_2 nanoparticles	Bacterial decrease from 4 to 6 Logs upon 90 min of exposure to simulated solar irradiation	[15]
Bacillus sp.	Water	Solar photocatalysis using Ag@SnO_2@ZnO core–shell nanocomposites	7 Log bacterial reduction within 210 min with a catalyst concentration of 500 mg/L	[64]
Escherichia coli	Wastewater	UV irradiation	Total inactivation after 60 min	[90]
Escherichia coli	Wastewater	Chlorination	Total inactivation after 120 min	[90]
Heterotrophic bacteria resistant to various antibiotics	Wastewater	Chlorination	Total inactivation	[89]

Table 3. *Cont.*

Antibiotic Resistance Genes (ARGs)	Aqueous Matrix	Treatment	Removal Level	Reference
blaTEM, ermB, qnrS, sulI and tetW	Wastewater	Wastewater treatment plant (WWTP)	Incomplete removal	[85]
tetA, tetB, tetE, tetG, tetH, tetS, tetT, tetX, sul1, sul2, qnrB and ermC	Wastewater	WWTP	Proliferation of ARGs through biological WWTP* processes	[87]
sul1, tetX, tetG and intI1	Municipal wastewater effluent	Chlorination	Reduction of ARGs in the range 1.20–1.49 Logs	[92]
sul1, tetX, tetG and intI1	Municipal wastewater effluent	UV/chlorination	Reduction of ARGs up to 2 Logs	[92]
mecA, ermB, sul1, tetA, tetW and tetX	Wastewater	WWTP	Incomplete removal	[93]
tetA, tetB, tetE, tetM, tetZ, tetW, sul1, sul2, sul3, gryA, qnrC, qnrD and parC	Wastewater	WWTP	Concentrations of the selected ARGs were kept relatively constant during treatment procedures	[14]
tetO, tetQ, tetW, tetH and tetZ	Wastewater	WWTP	Detectable ARGs in the effluents and possible proliferation	[86]
ereA, ereB, ermA, ermB, tetA, tetB, tetM and tetO	Wastewater	Chlorination	Limited removal levels (0.1–0.4 Logs)	[89]

6. Pilot-Scale Application

Although solar photocatalytic treatment of water and wastewater has successfully been tested in the laboratory, information regarding pilot- or large-scale applications is scarce (Tables 1 and 2). The pilot-scale applications that have been mainly tested are compound parabolic collectors (CPCs) and raceway ponds. Generally, both systems prove to be effective for the removal of persistent micro-contaminants of emerging concern and the elimination of waterborne pathogens.

Special key aspects for successful water treatment applications are the design and configuration of the photo-reactor. CPC solar reactors are one of the best approaches in order to enhance the efficacy of solar photocatalytic purification and disinfection of water [12]. These reactors are easy to use, cost-effective, and appropriate for point-of-use applications, since they can be constructed in various sizes. Raceway pond reactors were originally developed for micro-algal mass culture and are applied for the degradation of emerging micro-contaminants like pharmaceuticals and disinfection via solar photo-Fenton process [38,53]. Although they have less efficient optics than CPCs, they have a low construction cost and a large volume/surface ratio, which make them a quite competitive option for the treatment of secondary effluents [53]. Recent studies highlighted the prospect of scaling-up solar photocatalytic applications for water and wastewater treatment, considering those pilot-scale reactors as a post-secondary treatment step in WWTPs. This trend was followed by Barwal and Chaudhari, who designed and tested a hybrid bio-solar system with a moving bed biofilm reactor and a CPC for the purification and disinfection of municipal wastewaters [94].

Based on the above, large-scale applications of solar photocatalysis can serve as advanced tertiary treatment of wastewater and as an effective disinfection step in the water industry, especially in cases where other techniques are not suitable or feasible.

7. Future Perspectives

Although several AOPs have demonstrated supreme performance on water/wastewater treatment and disinfection over the last decades, solar photocatalysis is a relatively new area and there is lot yet to be explore and developed. The challenges are still numerous and many problems have to be overcome; however, the prospect of using solar light and energy combined with newly developed materials stands out as one sustainable alternative for environmental applications. Environmental protection and the economic cost are among the most important driving forces for the development of new methods that will be preserved and feasible in the course of time. In this respect, solar processes have all the characteristics and potential to be applied on a routine basis as efficient disinfection/decontamination treatment technologies. Also, they offer an ideal set-up for the synthesis of new, environmentally friendly materials that will serve as photocatalysts. Finally, the process scale-up, which has already begun, is a challenging task that will add to the overall science of water/wastewater treatment in an era where public health and environmental protection are the ultimate values for human beings. In a nutshell, water and wastewater purification and disinfection are listed among the topics that are balanced in the interface of science and engineering, and different disciplines must cooperate to deal with them successfully and constructively.

Author Contributions: Conceptualization—D.V., V.B. and D.M.; Investigation—D.V.; Writing—original draft preparation—D.V.; Writing—review and editing—D.M. and V.B. All authors have read and agreed to the published version of the manuscript.

Funding: This research received no external funding.

Conflicts of Interest: The authors declare no conflict of interest.

References

1. Zhang, Y.; Sivakumar, M.; Yang, S.; Enever, K.; Ramezanianpour, M. Application of solar energy in water treatment processes: A review. *Desalination* **2018**, *428*, 116–145. [CrossRef]
2. Dimitroula, H.; Daskalaki, V.M.; Frontistis, Z.; Kondarides, D.I.; Panagiotopoulou, P.; Xekoukoulotakis, N.P.; Mantzavinos, D. Solar photocatalysis for the abatement of emerging micro-contaminants in wastewater: Synthesis, characterization and testing of various TiO_2 samples. *Appl. Catal. B Environ.* **2012**, 283–291. [CrossRef]
3. Fagan, R.; McCormack, D.E.; Dionysiou, D.D.; Pillai, S.C. A review of solar and visible light active TiO_2 photocatalysis for treating bacteria, cyanotoxins and contaminants of emerging concern. *Mater. Sci. Semicond. Process.* **2016**, *42*, 2–14. [CrossRef]
4. Malato, S.; Fernández-Ibáñez, P.; Maldonado, M.; Blanco, J.; Gernjak, W. Decontamination and disinfection of water by solar photocatalysis: Recent overview and trends. *Catal. Today* **2009**, *147*, 1–59. [CrossRef]
5. Das, S.; Sinha, S.; Suar, M.; Yun, S.-I.; Mishra, A.; Tripathy, S.K. Solar-photocatalytic disinfection of Vibrio cholerae by using Ag@ZnO core–shell structure nanocomposites. *J. Photochem. Photobiol. B Biol.* **2015**, *142*, 68–76. [CrossRef]
6. Helali, S.; Polo-López, M.I.; Fernández-Ibáñez, P.; Ohtani, B.; Amano, F.; Malato, S.; Guillard, C. Solar photocatalysis: A green technology for E. coli contaminated water disinfection. Effect of concentration and different types of suspended catalyst. *J. Photochem. Photobiol. A Chem.* **2014**, *276*, 31–40. [CrossRef]
7. Aoudjit, L.; Martins, P.; Madjene, F.; Petrovykh, D.; Lanceros-Mendez, S. Photocatalytic reusable membranes for the effective degradation of tartrazine with a solar photoreactor. *J. Hazard. Mater.* **2018**, *344*, 408–416. [CrossRef] [PubMed]
8. Miralles-Cuevas, S.; Oller, I.; Pérez, J.S.; Malato, S. Removal of pharmaceuticals from MWTP effluent by nanofiltration and solar photo-Fenton using two different iron complexes at neutral pH. *Water Res.* **2014**, *64*, 23–31. [CrossRef]
9. Makropoulou, T.; Panagiotopoulou, P.; Venieri, D. N-doped TiO_2 photocatalysts for bacterial inactivation in water. *J. Chem. Technol. Biotechnol.* **2018**, *93*, 2518–2526. [CrossRef]
10. Abeledo-Lameiro, M.J.; Ares-Mazás, E.; Gómez-Couso, H. Evaluation of solar photocatalysis using TiO_2 slurry in the inactivation of Cryptosporidium parvum oocysts in water. *J. Photochem. Photobiol. B Biol.* **2016**, *163*, 92–99. [CrossRef]
11. Rizzo, L.; Della Sala, A.; Fiorentino, A.; Puma, G.L. Disinfection of urban wastewater by solar driven and UV lamp–TiO_2 photocatalysis: Effect on a multi drug resistant Escherichia coli strain. *Water Res.* **2014**, *53*, 145–152. [CrossRef] [PubMed]
12. Booshehri, A.Y.; Polo-Lopez, M.; Castro-Alférez, M.; He, P.; Xu, R.; Rong, W.; Malato, S.; Fernández-Ibáñez, P. Assessment of solar photocatalysis using Ag/BiVO 4 at pilot solar Compound Parabolic Collector for inactivation of pathogens in well water and secondary effluents. *Catal. Today* **2017**, *281*, 124–134. [CrossRef]
13. Carbajo, J.; Jiménez, M.; Miralles, S.; Malato, S.; Faraldos, M.; Bahamonde, A. Study of application of titania catalysts on solar photocatalysis: Influence of type of pollutants and water matrices. *Chem. Eng. J.* **2016**, *291*, 64–73. [CrossRef]

14. Xu, J.; Xu, Y.; Wang, H.; Guo, C.; Qiu, H.; He, Y.; Zhang, Y.; Li, X.; Meng, W. Occurrence of antibiotics and antibiotic resistance genes in a sewage treatment plant and its effluent-receiving river. *Chemosphere* **2015**, *119*, 1379–1385. [CrossRef]

15. Venieri, D.; Gounaki, I.; Bikouvaraki, M.; Binas, V.; Zachopoulos, A.; Kiriakidis, G.; Mantzavinos, D. Solar photocatalysis as disinfection technique: Inactivation of Klebsiella pneumoniae in sewage and investigation of changes in antibiotic resistance profile. *J. Environ. Manag.* **2017**, *195*, 140–147. [CrossRef]

16. Bouki, C.; Venieri, D.; Diamadopoulos, E. Detection and fate of antibiotic resistant bacteria in wastewater treatment plants: A review. *Ecotoxicol. Environ. Saf.* **2013**, *91*, 1–9. [CrossRef]

17. Comninellis, C.; Kapalka, A.; Malato, S.; Parsons, S.A.; Poulios, I.; Mantzavinos, D. Advanced oxidation processes for water treatment: Advances and trends for R&D. *J. Chem. Technol. Biotechnol.* **2008**, *83*, 769–776. [CrossRef]

18. Carp, O. Photoinduced reactivity of titanium dioxide. *Prog. Solid State Chem.* **2004**, *32*, 33–177. [CrossRef]

19. Venieri, D.; Fraggedaki, A.; Kostadima, M.; Chatzisymeon, E.; Binas, V.; Zachopoulos, A.; Kiriakidis, G.; Mantzavinos, D. Solar light and metal-doped TiO_2 to eliminate water-transmitted bacterial pathogens: Photocatalyst characterization and disinfection performance. *Appl. Catal. B Environ.* **2014**, *154-155*, 93–101. [CrossRef]

20. Khan, S.J.; Reed, R.H.; Rasul, M.G. Thin-film fixed-bed reactor for solar photocatalytic inactivation of Aeromonas hydrophila: Influence of water quality. *BMC Microbiol.* **2012**, *12*, 285. [CrossRef]

21. Fanourgiakis, S.; Frontistis, Z.; Chatzisymeon, E.; Venieri, D.; Mantzavinos, D. Simultaneous removal of estrogens and pathogens from secondary treated wastewater by solar photocatalytic treatment. *Glob. Nest J.* **2014**, *16*, 543–552.

22. Méndez-Arriaga, F.; Maldonado, M.I.; Gimenez, J.; Esplugas, S.; Malato, S. Abatement of ibuprofen by solar photocatalysis process: Enhancement and scale up. *Catal. Today* **2009**, *144*, 112–116. [CrossRef]

23. Devi, T.B.; Mohanta, D. Ahmaruzzaman Biomass derived activated carbon loaded silver nanoparticles: An effective nanocomposites for enhanced solar photocatalysis and antimicrobial activities. *J. Ind. Eng. Chem.* **2019**, *76*, 160–172. [CrossRef]

24. Kovacic, M.; Papac, J.; Kusic, H.; Karamanis, P.; Bozic, A.L. Degradation of polar and non-polar pharmaceutical pollutants in water by solar assisted photocatalysis using hydrothermal TiO_2-SnS2. *Chem. Eng. J.* **2020**, *382*, 122826. [CrossRef]

25. Peñas-Garzón, M.; Gómez-Avilés, A.; Belver, C.; Rodriguez, J.; Bedia, J. Degradation pathways of emerging contaminants using TiO_2-activated carbon heterostructures in aqueous solution under simulated solar light. *Chem. Eng. J.* **2020**, *392*, 124867. [CrossRef]

26. Salaeh, S.; Perisic, D.J.; Biosic, M.; Kusic, H.; Babic, S.; Stangar, U.L.; Dionysiou, D.D.; Bozic, A.L. Diclofenac removal by simulated solar assisted photocatalysis using TiO_2-based zeolite catalyst; mechanisms, pathways and environmental aspects. *Chem. Eng. J.* **2016**, *304*, 289–302. [CrossRef]

27. Peñas-Garzón, M.; Gómez-Avilés, A.; Bedia, J.; Rodriguez, J.J.; Belver, C. Effect of Activating Agent on the Properties of TiO_2/Activated Carbon Heterostructures for Solar Photocatalytic Degradation of Acetaminophen. *Materials* **2019**, *12*, 378. [CrossRef]

28. Akkari, M.; Aranda, P.; Belver, C.; Bedia, J.; Amara, A.B.H.; Ruiz-Hitzky, E. ZnO/sepiolite heterostructured materials for solar photocatalytic degradation of pharmaceuticals in wastewater. *Appl. Clay Sci.* **2018**, *156*, 104–109. [CrossRef]

29. Foteinis, S.; Monteagudo, J.M.; Durán, A.; Chatzisymeon, E. Environmental sustainability of the solar photo-Fenton process for wastewater treatment and pharmaceuticals mineralization at semi-industrial scale. *Sci. Total Environ.* **2018**, *612*, 605–612. [CrossRef]

30. Expósito, A.J.; Durán, A.; Monteagudo, J.M.; Acevedo, A.; Durán, A. Solar photo-degradation of a pharmaceutical wastewater effluent in a semi-industrial autonomous plant. *Chemosphere* **2016**, *150*, 254–257. [CrossRef]

31. Sirtori, C.; Zapata, A.; Oller, I.; Gernjak, W.; Agüera, A.; Malato, S. Decontamination industrial pharmaceutical wastewater by combining solar photo-Fenton and biological treatment. *Water Res.* **2009**, *43*, 661–668. [CrossRef] [PubMed]

32. Márquez, G.; Rodríguez, E.; Maldonado, M.I.; Álvarez, P.M. Integration of ozone and solar TiO_2-photocatalytic oxidation for the degradation of selected pharmaceutical compounds in water and wastewater. *Sep. Purif. Technol.* **2014**, *136*, 18–26. [CrossRef]

33. Gimeno, O.; García-Araya, J.; Beltrán, F.; Rivas, F.; Espejo, A. Removal of emerging contaminants from a primary effluent of municipal wastewater by means of sequential biological degradation-solar photocatalytic oxidation processes. *Chem. Eng. J.* **2016**, *290*, 12–20. [CrossRef]

34. Klamerth, N.; Rizzo, L.; Malato, S.; Maldonado, M.I.; Aguera, A.; Fernandezalba, A.R. Degradation of fifteen emerging contaminants at μgL−1 initial concentrations by mild solar photo-Fenton in MWTP effluents. *Water Res.* **2010**, *44*, 545–554. [CrossRef] [PubMed]

35. Klamerth, N.; Malato, S.; Agüera, A.; Fernández-Alba, A. Photo-Fenton and modified photo-Fenton at neutral pH for the treatment of emerging contaminants in wastewater treatment plant effluents: A comparison. *Water Res.* **2013**, *47*, 833–840. [CrossRef] [PubMed]

36. De La Cruz, N.; Dantas, R.F.; Giménez, J.; Esplugas, S. Photolysis and TiO$_2$ photocatalysis of the pharmaceutical propranolol: Solar and artificial light. *Appl. Catal. B Environ.* **2013**, *130*, 249–256. [CrossRef]

37. Jiménez-Tototzintle, M.; Oller, I.; Hernández-Ramírez, A.; Malato, S.; Maldonado, M. Remediation of agro-food industry effluents by biotreatment combined with supported TiO$_2$/H2O2 solar photocatalysis. *Chem. Eng. J.* **2015**, *273*, 205–213. [CrossRef]

38. Soriano-Molina, P.; Plaza-Bolaños, P.; Lorenzo, A.; Agüera, A.; Sánchez, J.G.; Malato, S.; Pérez, J.S. Assessment of solar raceway pond reactors for removal of contaminants of emerging concern by photo-Fenton at circumneutral pH from very different municipal wastewater effluents. *Chem. Eng. J.* **2019**, *366*, 141–149. [CrossRef]

39. Mecha, A.C.; Onyango, M.S.; Ochieng, A.; Momba, M.N.B. UV and solar photocatalytic disinfection of municipal wastewater: Inactivation, reactivation and regrowth of bacterial pathogens. *Int. J. Environ. Sci. Technol.* **2019**, *16*, 3687–3696. [CrossRef]

40. Levchuk, I.; Homola, T.; Moreno-Andrés, J.; Rueda-Márquez, J.J.; Dzik, P.; Moríñigo, M.Á.; Sillanpää, M.; Manzano, M.A.; Vahala, R. Solar photocatalytic disinfection using ink-jet printed composite TiO$_2$/SiO2 thin films on flexible substrate: Applicability to drinking and marine water. *Sol. Energy* **2019**, *191*, 518–529. [CrossRef]

41. Venieri, D.; Tournas, F.; Gounaki, I.; Binas, V.; Zachopoulos, A.; Kiriakidis, G.; Mantzavinos, D. Inactivation ofStaphylococcus aureusin water by means of solar photocatalysis using metal doped TiO$_2$ semiconductors. *J. Chem. Technol. Biotechnol.* **2017**, *92*, 43–51. [CrossRef]

42. Sreeja, S.; K Shetty, V. Photocatalytic water disinfection under solar irradiation by Ag@TiO$_2$ core-shell structured nanoparticles. *Sol. Energy* **2017**, *157*, 236–243. [CrossRef]

43. Afsharnia, M.; Kianmehr, M.; Biglari, H.; Dargahi, A.; Karimi, A. Disinfection of dairy wastewater effluent through solar photocatalysis processes. *Water Sci. Eng.* **2018**, *11*, 214–219. [CrossRef]

44. Barreca, S.; Vélez-Colmenares, J.; Pace, A.; Orecchio, S.; Pulgarin, C. Escherichia coli inactivation by neutral solar heterogeneous photo-Fenton (HPF) over hybrid iron/montmorillonite/alginate beads. *J. Environ. Chem. Eng.* **2015**, *3*, 317–324. [CrossRef]

45. Sichel, C.; De Cara, M.; Tello, J.; Blanco, J.; Ibáñez, P.F. Solar photocatalytic disinfection of agricultural pathogenic fungi: Fusarium species. *Appl. Catal. B Environ.* **2007**, *74*, 152–160. [CrossRef]

46. Misstear, D.B.; Gill, L.W. The inactivation of phages MS2, ΦX174 and PR772 using UV and solar photocatalysis. *J. Photochem. Photobiol. B Biol.* **2012**, *107*, 1–8. [CrossRef]

47. Venieri, D.; Gounaki, I.; Binas, V.; Zachopoulos, A.; Kiriakidis, G.; Mantzavinos, D. Inactivation of MS2 coliphage in sewage by solar photocatalysis using metal-doped TiO$_2$. *Appl. Catal. B Environ.* **2015**, *178*, 54–64. [CrossRef]

48. Nieto-Juarez, J.I.; Kohn, T. Virus removal and inactivation by iron (hydr)oxide-mediated Fenton-like processes under sunlight and in the dark. *Photochem. Photobiol. Sci.* **2013**, *12*, 1596–1605. [CrossRef]

49. Waso, M.; Khan, S.; Singh, A.; McMichael, S.; Ahmed, W.; Fernández-Ibáñez, P.; Byrne, J.; Khan, W. Predatory bacteria in combination with solar disinfection and solar photocatalysis for the treatment of rainwater. *Water Res.* **2020**, *169*, 115281. [CrossRef]

50. Sichel, C.; Blanco, J.; Malato, S.; Ibáñez, P.F. Effects of experimental conditions on E. coli survival during solar photocatalytic water disinfection. *J. Photochem. Photobiol. A Chem.* **2007**, *189*, 239–246. [CrossRef]

51. Gomes, A.I.; Santos, J.C.; Vilar, V.J.; Boaventura, R.A. Inactivation of Bacteria E. coli and photodegradation of humic acids using natural sunlight. *Appl. Catal. B Environ.* **2009**, *88*, 283–291. [CrossRef]

52. García-Fernández, I.; Fernández-Calderero, I.; Polo-López, M.I.; Fernández-Ibáñez, P. Disinfection of urban effluents using solar TiO_2 photocatalysis: A study of significance of dissolved oxygen, temperature, type of microorganism and water matrix. *Catal. Today* **2015**, *240*, 30–38. [CrossRef]

53. De la Obra Jiménez, I.; Esteban García, B.; Rivas Ibáñez, G.; Casas López, J.L.; Sanchez Pérez, J.A. Continuous flow disinfection of WWTP secondary effluents by solar photo-Fenton at neutral pH in raceway pond reactors at pilot plant scale. *Appl. Catal. B Environ.* **2019**, *247*, 115–123. [CrossRef]

54. Venieri, D.; Chatzisymeon, E.; Gonzalo, M.S.; Rosal, R.; Mantzavinos, D. Inactivation of Enterococcus faecalis by TiO_2-mediated UV and solar irradiation in water and wastewater: Culture techniques never say the whole truth. *Photochem. Photobiol. Sci.* **2011**, *10*, 1744. [CrossRef]

55. Karunakaran, C.; Vijayabalan, A.; Manikandan, G.; Gomathisankar, P. Visible light photocatalytic disinfection of bacteria by Cd–TiO_2. *Catal. Commun.* **2011**, *12*, 826–829. [CrossRef]

56. Malato, S.; Maldonado, M.I.; Fernández-Ibáñez, P.; Oller, I.; Polo, I.; Sánchez-Moreno, R. Decontamination and disinfection of water by solar photocatalysis: The pilot plants of the Plataforma Solar de Almeria. *Mater. Sci. Semicond. Process.* **2016**, *42*, 15–23. [CrossRef]

57. Pelaez, M.; Nolan, N.T.; Pillai, S.C.; Seery, M.K.; Falaras, P.; Kontos, A.G.; Dunlop, P.S.; Hamilton, J.W.; Byrne, J.; O'Shea, K.; et al. A review on the visible light active titanium dioxide photocatalysts for environmental applications. *Appl. Catal. B Environ.* **2012**, *125*, 331–349. [CrossRef]

58. Binas, V.; Venieri, D.; Kotzias, D.; Kiriakidis, G. Modified TiO_2 based photocatalysts for improved air and health quality. *J. Mater.* **2017**, *3*, 3–16. [CrossRef]

59. Papadimitriou, V.C.; Stefanopoulos, V.G.; Romanias, M.N.; Papagiannakopoulos, P.; Sambani, K.; Tudose, V.; Kiriakidis, G. Determination of photo-catalytic activity of un-doped and Mn-doped TiO_2 anatase powders on acetaldehyde under UV and visible light. *Thin Solid Films* **2011**, *520*, 1195–1201. [CrossRef]

60. Byrne, J.A.; Fernandez-Ibañez, P.A.; Dunlop, P.S.M.; Alrousan, D.M.A.; Hamilton, J.W.J. Photocatalytic Enhancement for Solar Disinfection of Water: A Review. *Int. J. Photoenergy* **2011**, *2011*, 1–12. [CrossRef]

61. Ede, S.; Hafner, L.; Dunlop, P.; Byrne, J.; Will, G. Photocatalytic Disinfection of Bacterial Pollutants Using Suspended and Immobilized TiO_2 Powders. *Photochem. Photobiol.* **2012**, *88*, 728–735. [CrossRef] [PubMed]

62. Dunlop, P.; McMurray, T.A.; Hamilton, J.W.J.; Byrne, J.A. Photocatalytic inactivation of Clostridium perfringens spores on TiO_2 electrodes. *J. Photochem. Photobiol. A Chem.* **2008**, *196*, 113–119. [CrossRef]

63. Das, S.; Sinha, S.; Das, B.; Jayabalan, R.; Suar, M.; Mishra, A.; Tamhankar, A.J.; Lundborg, C.S.; Tripathy, S.K. Disinfection of Multidrug Resistant Escherichia coli by Solar-Photocatalysis using Fe-doped ZnO Nanoparticles. *Sci. Rep.* **2017**, *7*, 104. [CrossRef] [PubMed]

64. Das, S.; JyotiMisraab, A.; Rahmanab, A.; Dasc, B.; Jayabalan, R.J.; Tamhankarbd, A.; Mishrab, A.; Lundborg, C.S.; Tripathy, S.K. Ag@SnO2@ZnO core-shell nanocomposites assisted solar-photocatalysis downregulates multidrug resistance in Bacillus sp.: A catalytic approach to impede antibiotic resistance. *Appl. Catal. B Environ.* **2019**, *259*, 118065. [CrossRef]

65. Ayodhya, D.; Venkatesham, M.; Kumari, A.S.; Reddy, G.B.; Guttena, V. One-pot sonochemical synthesis of CdS nanoparticles: Photocatalytic and electrical properties. *Int. J. Ind. Chem.* **2015**, *6*, 261–271. [CrossRef]

66. Yi, Z.; Ye, J.; Kikugawa, N.; Kako, T.; Ouyang, S.; Stuart-Williams, H.; Yang, H.; Cao, J.; Luo, W.; Li, Z.; et al. An orthophosphate semiconductor with photooxidation properties under visible-light irradiation. *Nat. Mater.* **2010**, *9*, 559–564. [CrossRef]

67. Eswar, N.K.; Ramamurthy, P.C.; Madras, G. Enhanced sunlight photocatalytic activity of Ag_3PO_4 decorated novel combustion synthesis derived TiO_2nanobelts for dye and bacterial degradation. *Photochem. Photobiol. Sci.* **2015**, *14*, 1227–1237. [CrossRef]

68. Xu, J.-W.; Gao, Z.-D.; Han, K.; Liu, Y.; Song, Y.-Y. Synthesis of Magnetically Separable Ag_3PO_4/TiO_2/Fe_3O_4 Heterostructure with Enhanced Photocatalytic Performance under Visible Light for Photoinactivation of Bacteria. *Appl. Mater. Interfaces* **2014**, *6*, 15122–15131. [CrossRef]

69. Yang, X.; Qin, J.; Jiang, Y.; Li, R.; Li, Y.; Tang, H. Bifunctional TiO_2/Ag_3PO_4/graphene composites with superior visible light photocatalytic performance and synergistic inactivation of bacteria. *RSC Adv.* **2014**, *4*, 18627–18636. [CrossRef]

70. Springer-Verlag GmbH Germany. Advances in Photocatalytic Disinfection. In *Frontiers of Green Catalytic Selective Oxidations*; Springer-Verlag GmbH Germany: Heidelberg, Germany, 2017; pp. 1–16. Available online: https://link.springer.com/book/10.1007/978-3-662-53496-0 (accessed on 30 November 2020).

71. IWA Publishing 2nd ed. Wastewater and Biosolids Management. Available online: https://www.iwapublishing.com/books/9781789061659/wastewater-and-biosolids-management-2nd-edition (accessed on 28 May 2020).

72. Ioannidou, E.; Frontistis, Z.; Antonopoulou, M.; Venieri, D.; Konstantinou, I.K.; Kondarides, D.I.; Mantzavinos, D. Solar photocatalytic degradation of sulfamethoxazole over tungsten–Modified TiO_2. *Chem. Eng. J.* **2017**, *318*, 143–152. [CrossRef]

73. Outsiou, A.; Frontistis, Z.; Ribeiro, R.S.; Antonopoulou, M.; Konstantinou, I.; Silva, A.M.T.; Faria, J.L.; Gomes, H.T.; Mantzavinos, D. Activation of sodium persulfate by magnetic carbon xerogels (CX/CoFe) for the oxidation of bisphenol A: Process variables effects, matrix effects and reaction pathways. *Water Res.* **2017**, *124*, 97–107. [CrossRef] [PubMed]

74. Pepper, I.; Gerba, C.; Gentry, T. *Environmental Microbiology*; Elsevier: Amsterdam, The Netherlands, 2015; ISBN 9780123946263.

75. Zuo, X.; Hu, J.; Chen, M. The role and fate of inorganic nitrogen species during UVA/TiO_2 disinfection. *Water Res.* **2015**, *80*, 12–19. [CrossRef] [PubMed]

76. Marugán, J.; Van Grieken, R.; Pablos, C. Kinetics and influence of water composition on photocatalytic disinfection and photocatalytic oxidation of pollutants. *Environ. Technol.* **2010**, *31*, 1435–1440. [CrossRef] [PubMed]

77. Xiao, G.; Zhang, X.; Zhang, W.; Zhang, S.; Su, H.; Tan, T. Visible-light-mediated synergistic photocatalytic antimicrobial effects and mechanism of Ag-nanoparticles@chitosan–TiO_2 organic–inorganic composites for water disinfection. *Appl. Catal. B Environ.* **2015**, *170–171*, 255–262. [CrossRef]

78. Adefisoye, M.A.; Nwodo, U.U.; Green, E.; Okoh, A.I. Quantitative PCR Detection and Characterisation of Human Adenovirus, Rotavirus and Hepatitis A Virus in Discharged Effluents of Two Wastewater Treatment Facilities in the Eastern Cape, South Africa. *Food Environ. Virol.* **2016**, *8*, 262–274. [CrossRef] [PubMed]

79. Jebri, S.; Hmaied, F.; Jofre, J.; Yahya, M.; Mendez, J.; Barkallah, I.; Hamdi, M. Effect of gamma irradiation on bacteriophages used as viral indicators. *Water Res.* **2013**, *47*, 3673–3678. [CrossRef]

80. Ditta, I.B.; Steele, A.; Liptrot, C.; Tobin, J.; Tyler, H.; Yates, H.M.; Sheel, D.W.; Foster, H.A. Photocatalytic antimicrobial activity of thin surface films of TiO_2, CuO and TiO_2/CuO dual layers on Escherichia coli and bacteriophage T4. *Appl. Microbiol. Biotechnol.* **2008**, *79*, 127–133. [CrossRef]

81. Dalrymple, O.K.; Stefanakos, E.; Trotz, M.A.; Goswami, D.Y. A review of the mechanisms and modeling of photocatalytic disinfection. *Appl. Catal. B Environ.* **2010**, *98*, 27–38. [CrossRef]

82. Süß, J.; Volz, S.; Obst, U.; Schwartz, T. Application of a molecular biology concept for the detection of DNA damage and repair during UV disinfection. *Water Res.* **2009**, *43*, 3705–3716. [CrossRef]

83. Rochelle, P.A.; Upton, S.J.; Montelone, B.A.; Woods, K. The response of Cryptosporidium parvum to UV light. *Trends Parasitol.* **2005**, *21*, 81–87. [CrossRef]

84. Manaia, C.M.; Macedo, G.; Fatta-Kassinos, D.; Nunes, O.C. Antibiotic resistance in urban aquatic environments: Can it be controlled? *Appl. Microbiol. Biotechnol.* **2016**, *100*, 1543–1557. [CrossRef] [PubMed]

85. Rodriguez-Mozaz, S.; Chamorro, S.; Marti, E.; Huerta, B.; Gros, M.; Sànchez-Melsió, A.; Borrego, C.M.; Barceló, J.; Balcázar, J.L. Occurrence of antibiotics and antibiotic resistance genes in hospital and urban wastewaters and their impact on the receiving river. *Water Res.* **2015**, *69*, 234–242. [CrossRef] [PubMed]

86. Al-Jassim, N.; Ansari, M.I.; Harb, M.; Hong, P.-Y. Removal of bacterial contaminants and antibiotic resistance genes by conventional wastewater treatment processes in Saudi Arabia: Is the treated wastewater safe to reuse for agricultural irrigation? *Water Res.* **2015**, *73*, 277–290. [CrossRef] [PubMed]

87. Mao, D.; Yu, S.; Rysz, M.; Luo, Y.; Yang, F.; Li, F.; Hou, J.; Mu, Q.; Alvarez, P.J. Prevalence and proliferation of antibiotic resistance genes in two municipal wastewater treatment plants. *Water Res.* **2015**, *85*, 458–466. [CrossRef] [PubMed]

88. Jahne, M.A.; Rogers, S.W.; Ramler, I.P.; Holder, E.; Hayes, G. Hierarchal clustering yields insight into multidrug-resistant bacteria isolated from a cattle feedlot wastewater treatment system. *Environ. Monit. Assess.* **2015**, *187*, 1–15. [CrossRef] [PubMed]

89. Yuan, Q.-B.; Guo, M.-T.; Yang, J. Fate of Antibiotic Resistant Bacteria and Genes during Wastewater Chlorination: Implication for Antibiotic Resistance Control. *PLoS ONE* **2015**, *10*, e0119403. [CrossRef] [PubMed]

90. Rizzo, L.; Fiorentino, A.; Anselmo, A. Advanced treatment of urban wastewater by UV radiation: Effect on antibiotics and antibiotic-resistant E. coli strains. *Chemosphere* **2013**, *92*, 171–176. [CrossRef]

91. Rizzo, L.V.; Sannino, D.; Vaiano, V.; Sacco, O.; Scarpa, A.; Pietrogiacomi, D. Effect of solar simulated N-doped TiO$_2$ photocatalysis on the inactivation and antibiotic resistance of an E. coli strain in biologically treated urban wastewater. *Appl. Catal. B Environ.* **2014**, *144*, 369–378. [CrossRef]

92. Zhang, Y.; Zhuang, Y.; Geng, J.; Ren, H.; Zhang, Y.; Ding, L.; Xu, K. Inactivation of antibiotic resistance genes in municipal wastewater effluent by chlorination and sequential UV/chlorination disinfection. *Sci. Total Environ.* **2015**, *512–513*, 125–132. [CrossRef]

93. Naquin, A.; Shrestha, A.; Sherpa, M.; Nathaniel, R.; Boopathy, R. Presence of antibiotic resistance genes in a sewage treatment plant in Thibodaux, Louisiana, USA. *Bioresour. Technol.* **2015**, *188*, 79–83. [CrossRef]

94. Barwal, A.; Chaudhary, R. Feasibility study for the treatment of municipal wastewater by using a hybrid bio-solar process. *J. Environ. Manag.* **2016**, *177*, 271–277. [CrossRef] [PubMed]

 *sustainability*

Review

Overview of the Policies for Phasing Out Ocean Dumping of Sewage Sludge in the Republic of Korea

Chang Soo Chung [1,2], Ki-Young Choi [1], Chang-Joon Kim [1], Jun-Mo Jung [1] and Yeon S. Chang [1,2,*]

[1]   Korea Institute of Ocean Science and Technology, 385, Haeyang-ro, Yeongdo-gu, Busan 49111, Korea; cschung@kiost.ac.kr (C.S.C.); kychoi@kiost.ac.kr (K.-Y.C.); kcj201@kiost.ac.kr (C.-J.K.); jungjm86@kiost.ac.kr (J.-M.J.)
[2]   KIOST School, University of Science and Technology (UST), 385, Haeyang-ro, Yeongdo-gu, Busan 49111, Korea
*   Correspondence: yeonschang@kiost.ac.kr; Tel.: +82-51-664-3566

Received: 6 May 2020; Accepted: 1 June 2020; Published: 3 June 2020

Abstract: Ocean dumping of municipal sewage sludge (MSS) that was treated in wastewater treatment plants in the Republic of Korea (ROK) began in 1993 and has sharply increased thereafter; this deteriorated the benthic environment of the dumping sites, consequently necessitating relevant policies to be developed to reduce dumping. This review introduces the outcomes of policies used to phase out ocean dumping of MSS in ROK and provides a method for improving contaminated environments. We first review a previous report submitted under the London Protocol in 2016 and then provide additional data collected since then. In addition, we introduce a scientific research result that reduced the concentration of harmful substances in the dumping sites by capping the dumping area. ROK established policies to phase out the dumping in 2006, which had immediate impacts, with dumping of MSS terminated in 2012. These policies were then expanded to terminate dumping of all types of sewage sludge in 2016, due to the fast and strict application of actions based on intergovernmental cooperation and societal consensus. In addition, the capping method that covered the contaminated sediments with dredged materials was effective. The success of the evaluated policies and research could be effectively applied to areas with similar circumstances.

Keywords: ocean dumping; sewage sludge; capping method; London Protocol

1. Introduction

Sewage sludge is an end-product of primary and secondary treatments of sewage effluent arising from urban sources and may contain high concentrations of heavy metals and persistent synthetic organic compounds, depending on its origin [1–4]. Consequently, careless dumping of municipal sewage sludge (MSS) to the sea has resulted in serious marine pollution and has raised concerns about the damage to the marine environment and potential safety threats [5,6]. In addition, there has been growing awareness that instead of disposal of sewage sludge, it can be recycled.

With the increasing demand for international regulation, the practice of ocean dumping of sewage sludge has been banned in Europe since 1998 by the Convention for the Protection of the Marine Environment of the North–East Atlantic, and the London Protocol (hereafter, LP) came into effect in 2006. The LP was organized to protect the marine environment from human activities under the International Maritime Organization (IMO), and it has focused on a precautionary approach to environmental protection from the dumping of waste or other materials [7,8]. The Republic of Korea (hereafter, ROK) joined the London Convention (hereafter, LC) on December 21, 1993, and the LP that was agreed in 1996 to further modernize and eventually to replace the LC, on January 22, 2009.

The Ministry of Ocean and Fisheries (hereafter, MOF) of ROK administers the ocean dumping of wastes, including MSS, under the Marine Environmental Management Act to meet the international obligations under the LP. In pursuing the precautionary principles of the LP, the MOF recognized that using MSS as a resource instead of just dumping it into the sea could be a more effective approach. Therefore, it has aggressively promoted the policy with efforts to minimize conflicts with stakeholders.

In July 2016, the Korea Institute of Ocean Science and Technology (hereafter, KIOST) of ROK submitted a report to the office of the London Convention and London Protocol (hereafter, LC/LP) with the title of "Barriers to compliance with the termination of the ocean dumping of sewage sludge in the Republic of Korea" [9]. In the report, ROK introduced a legal system and background for phasing out the ocean dumping of MSS before the establishment of the policy in 2012. Furthermore, it introduced the progress of the policy on the establishment of annual reduction goals, action levels to improve the efficiency of the policy, a surveillance system to observe illegal dumping, monitoring outcomes, management plans, and public activities with the cooperation of governmental agencies.

A study of similar analysis can be found in Swanson et al. [10], in which the lessons learned from the practice of ocean dumping of sewage sludge in New York City since the early 20th century was reviewed to provide understanding of the long-term effect of policies that had been piecemeal and changed according to shifting circumstances. Ocean dumping of sewage began in New York City in 1924 at the 12–Mile Site, and it moved to the 106–Mile Site in 1986. In response to political pressure, the practice of ocean dumping ended entirely in 1992. The study revealed that, although it was cheaper than other disposal methods, ocean dumping of sewage sludge could deteriorate the benthic environment in a wide area around the dumping site. In addition, Ofira [11] introduced efforts and outcomes to clean up the contaminated beaches along the coastlines of New York and New Jersey as a result of the NY Bight Restoration Plan. The results suggested that although progress was made successfully, policy directives may need to be refined based on the current situation. For example, initial objectives to ensure that no beaches are closed during the summer season may not be achievable based on the research data on environmental contamination, and thus an amendment to these policies may be required.

Similarly, the objective of this review is to examine the policies and outcomes concerning phasing out the ocean dumping of sewage sludge in ROK and to share its experience with countries in overcoming barriers to LP compliance in tandem with a previous report submitted to the office of the LP [9]. We provide additional data on the outcomes of the phasing out policy and compare the statistics between the ocean dumping and landfills of MSS to show the efficiency of the administrative actions. In addition, follow-up measures taken after 2016 are introduced, including extended monitoring and management results. This study's findings can provide useful information for the future development of long-term strategies and policies for minimizing ocean dumping and protecting the marine environment in coastal seas. Specifically, we suggest a method for improving the contaminated benthic environment at a dumping site based on statistical results, with specific conditions on where direct dredging was not feasible because of the distance from the shore and the water depth; this approach may be effectively applied to areas with similar circumstances.

2. Literature/Policy Review

2.1. Summary of 2016 Report to Office of LP

In this section, the previous report submitted to the office of the LP in 2016 [9] is summarized by briefly introducing the progress and outcome of the MOF's administrative policy to phase out the ocean dumping of sewage sludge in the ROK up to 2016. This summary is necessary for readers to understand the background of the policy and its progress, as the previous report is not easily accessible. Furthermore, the summary can be used to understand the additional statistics and measures provided in this review.

Background

The ROK began ocean dumping in 1993 to reduce the load of waste produced on land and to prevent excessive degradation of river and coastal environments [12]. Consequently, the Waste Management Act pronounced a ban on the direct landfilling of MSS in 1997 and went into effect in 2006. However, these legal measures caused substantial amounts of MSS to be diverted from land filling to ocean dumping. The dumping of MSS to the sea commenced in 1993, with a sharp build-up in quantity from approximately 10×10^3 m^3 to 8812×10^3 m^3 in 2006.

As a consequence of this increased influx of MSS, substantial enrichments were observed for trace metals including Cr, Cd, and Pb in the surficial dump site sediments, and this has affected the distributional patterns of benthic species in dump sites [13–17]. The habitats at these dump sites have been negatively impacted by anthropogenic activities, resulting in considerable changes to the structure of these ecosystems. This led to widespread awareness of environmental damage caused by the ocean dumping of MSS that evolved into a major social crisis, producing intense social conflicts, which provided a key motivation for establishing the policy to phase out ocean dumping of MSS [12].

Internationally, a common trend has been to enact waste treatment policies in the order of most-to-least impactful on the environment by allowing ocean dumping only when inland treatment/disposal methods are not available. In this context, most contracting parties of the LP had banned the ocean dumping of MSS prior to the year 2000, as the LP requires that the hierarchy of waste management options be followed according to increasing environmental impact; this reality encouraged the MOF to prepare relevant policies and regulations concerning marine environment protection.

2.2. Progress of the Phase-Out Policy

2.2.1. Annual Reduction Goal

MOF prepared the implementation of the LP in the ROK by arranging regulations for ocean dumping of MSS in 2006; specifically, the MOF promulgated a revised enforcement rule for the Marine Environment Management Act. The MOF then established the Master Plan of Ocean Dumping Management, and it was approved by the Cabinet Council in March 2006. The plan focused on strengthening the restriction of ocean dumping of wastes by establishing a principle of land-based priority treatment and a scientific system of environmental management of dumping sites. This plan aimed to reduce the ocean dumping of MSS through inter-agency consultations and professional advice on the status of inland treatment/disposal. The plan also set a goal in 2006 to reduce the annual amount of MSS dumped into the sea by 50% by 2011, as listed in Table 1.

Table 1. Ministry of Ocean and Fisheries' stated goal to reduce the annual amount of municipal sewage sludge dumped to the sea.

Year	2006	2007	2008	2009	2010	2011
Goal amount (unit: $\times10^3$ m^3)	9000	8000	6000	5000	4500	4000
Amount dumped to the sea (unit: $\times10^3$ m^3)	8812	7451	6173	4785	4478	3972

To achieve this goal, MOF strengthened the standards of ocean dumping pollution by introducing the "polluter-pays principle", by which the party responsible for producing pollution is responsible for paying for the damage done to the natural environment. This policy is enforced by conducting monitoring at dump sites on a regular basis, monitoring illegal activities by dumping vessels by strengthening cooperation with other organizations, and by building consensus among stakeholders. Consequently, the amount of ocean dumping of MSS has continuously decreased by more than 10% per year and the actual amount of ocean dumping in 2011 just before the ban of ocean dumping of MSS in 2012 was 3.97 million m^3, which was less than the targeted value (4 million m^3). Since 2012, the amount has been zero until 2019.

2.2.2. Action Levels

The maximum permissible contents of contaminants that can be dumped into the sea have been set as action levels by the LP (IMO) waste assessment guidelines [7]). The action levels have effectively served as policy measures that have assisted in reducing the demand for ocean dumping of MSS, and pollution loads, and for mitigating pollution at dumping sites. Furthermore, these measures have been useful in reducing the amount of waste by converting it to recycling materials. In Table 2, the rates of recycling, landfill, incineration and ocean dumping out of the total mass of sewage sludge is compared between 2005 and 2012, which shows that energy efficiency was greatly enhanced due to the reduction in ocean dumping.

Table 2. Rate of recycling, landfill, incineration and ocean dumping out of total mass of sewage sludge in Korea between 2005 and 2012 [18].

Year	Recycle (%)	Landfill (%)	Incineration (%)	Ocean Dumping (%)
2005	4.8	1.7	11.2	78
2012	43.0	15.0	35.0	0

Based on the LP's guidelines, the MOF also developed its own action levels that provide criteria for permissible contaminants by analyzing sewage sludge in more than one hundred sewage treatment plants in the ROK (Table 2). The upper and lower limits of the permissible contents of the harmful substances in the list have served as a useful basis to determine how much of those substances could be safely contained in the ocean (Table 3). Using the action levels based on LP guidelines, the wastes were categorized into three groups before deciding their suitability for ocean dumping as follows:

1. Upper level: wastes that exceed maximum levels of permissible contents shall not be dumped to avoid acute or chronic effects on human health or impacts on sensitive marine organisms that are representative of the marine ecosystem.
2. Lower level: wastes that are below minimal levels of permissible contents are of little environmental concern in relation to dumping.
3. Wastes that are in between the lower and upper limits require more detailed assessments such as toxicity tests before determining their suitability for dumping.

Contractor companies should improve their capacity to purify sewage sludge to satisfy these criteria. Following international standards, luminous bacteria and benthic amphipods were used for toxicity tests, and only the MSS that passed the test criteria has been permitted for ocean dumping since 2008. In case of failure, the sludge should be treated or disposed of on land. The failure rates in the tests increased and became similar to the rate of reduction in the amount of MSS dumped in the ocean, and its amount in 2011 became half of that in 2005.

Table 3. Ocean dumping criteria of sewage sludge in the Republic of Korea.

Substances	Unit	Upper Level	Lower Level
Mineral oil		10,000	2000
CN		200	40
Phenol		4000	800
Cr		1850	370
Zn		9000	1800
Cu	mg/kg dry weight	2000	400
Cd		20	4
Hg		5	1
Organic phosphorus		100	20
As		145	29
Pb		1100	220
Polychlorinated biphenyl (PCB)-28		0.15	0.03
PCB-52		0.15	0.03
PCB-101		0.15	0.03
PCB-118	µg/kg dry weight	0.15	0.03
PCB-138		0.15	0.03
PCB-153		0.15	0.03
PCB-180		0.15	0.03
Naphthalene		4	0.8
Phenanthrene		5	1
Anthracene		4	0.8
Benzo[a]pyrene	µg/kg dry weight	4.5	0.9
Fluoranthene		10	2.5
Benzo[a]anthracene		5	1
Benzo[b]fluoranthene		4	0.8

2.2.3. Polluter-Pays Approach

The main reason that ocean dumping of sewage sludge was preferred to inland treatment/disposal methods was the minimal cost [10]. Therefore, the amount of ocean dumping can be controlled if the responsibility of the dumping party is increased in case of pollution; this summarizes the practical application of the LP's "polluter-pays principle". Perceiving its efficiency, the MOF also applied the principle by amending the Marine Environment Management Act to impose charges on ocean dumping of sewage sludge starting in 2006. The charge would increase if the dumping sewage sludge contained high concentrations of contaminants, and the imposed charges were deposited to the fisheries development fund that is used for marine environment improvement.

2.3. Surveillance

Occasionally, illegal dumping would occur, as the sewage sludge was disposed outside the designated sites to reduce the transportation cost. For example, a total of 188 cases were detected that violated ocean dumping standards from 2005 to 2007 in the ROK. To inspect such violations, MOF developed the automatic identification system (AIS) by which the navigation status of waste transport vessels could be tracked using GPS. The illegal dumping of sewage sludge could be detected by the Korean Coast Guard by monitoring the locations and traces of the vessels, and all waste transport vessels have been equipped with AIS since 2008 to prevent illegal practices.

In accordance with this surveillance regime, appropriate monitoring of waste dump sites has become an important approach, because environmental impacts from ocean dumping of sewage sludge are severe and long-lasting [19]; moreover, the LP requires monitoring of waste dump sites as a precautionary approach. Since 2004, KIOST has conducted annual monitoring at designated dumping sites in which their environmental conditions are evaluated and remedies are proposed if problems are detected. The results from the monitoring have been utilized to minimize damages to the marine environment through the construction of regulations on waste dumping in these sea areas, with the ultimate goal of restoring the contaminated areas. Since 2006, for example, ocean dumping activities have been banned in some sectors of dump sites that have suffered from concentrated contamination; accordingly, they were designated for restoration based on KIOST's monitoring results [16].

2.4. Intergovernmental Coordination and Consensus Building

Until 2005, the amount of ocean dumping of MSS in the ROK rapidly increased, and the most common reason for this was the enhanced regulatory control on inland waste treatment/disposal methods that left ocean dumping as the only viable option. For instance, the Ministry of Environment of ROK announced a ban on the landfill of sewage sludge in 1997, which had resulted in a sharp increase in the ocean dumping of MSS from 1996 until 2005, when the ban came into effect. Figure 1 compares the annual variation of the MSS amounts dumped in the ocean and sent to landfills. The amount of landfilled sewage sludge sharply decreased from 1993–2005, which was due to the banning of landfills by the Waste Management Act as pronounced in 1997 and implemented in 2006.

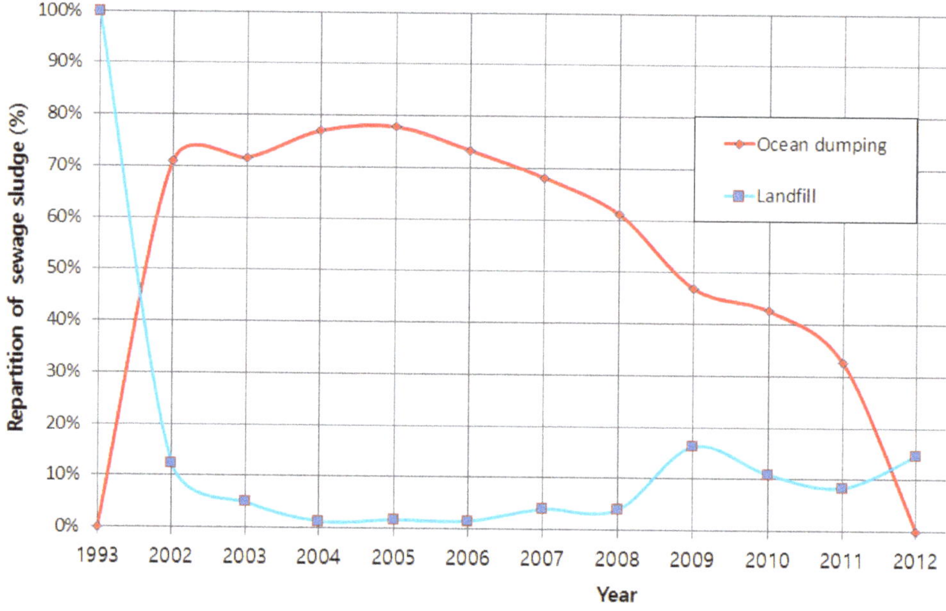

Figure 1. Comparison of the amount of municipal sewage sludge (MSS) disposed of via ocean dumping (blue) and landfill (red) from 1993, when ocean dumping commenced, to 2012, when ocean dumping was terminated. The amounts are normalized by the landfill amount in 1993. The figure is modified from the original version in which only a repartition of ocean dumping of sewage sludge was plotted [9]. Data was not available from 1994 to 2001.

Thus, to reduce the ocean dumping of land-based wastes, cooperation was necessary between corresponding governmental agencies in reducing in both land- and ocean-based wastes. The MOF started a task force team to collect data on the amount of MSS produced in each local district and their present treatment conditions to establish future plans through cooperation with other governmental agencies and by conducting field studies at the treatment facilities. In 2006, the Master Plan of Ocean Dumping Management was established and implemented based on the data collected by the team.

The consensus and active participation of the public and stakeholders is an important ingredient in the successful implementation of phasing out the ocean dumping of MSS. For this purpose, the MOF determined the crucial issues that various stakeholders faced and held public hearings, making efforts to ensure compromises between them. Combining the data collected from the surveys made for the inland treatment/disposal and ocean dumping practices of the sewage sludge, the MOF concluded that a reduction in ocean dumping and ultimately phasing out the practice while increasing the number of the inland disposal facilities would result in stronger benefits for the country. In addition, the MOF published professional opinions with the help of KIOST through mass media, and reached out to the public through on-site education to understand the impact of marine pollution caused by ocean dumping of sewage sludge and thus to enhance citizens' awareness.

2.5. Monitoring since 2016

Recognizing that the main reason for the increase in the ocean dumping of municipal sewage waste was the ban on landfills, the MOF closely cooperated with other government agencies and developed the Master Plan for Ocean Dumping Management in 2006. In addition, the MOF prepared regulations for the ROK to join the LP in 2009 so that its measures could be implemented nationally, which led to the establishment of the amendment to the Marine Environment Management Act in 2012. The effect of these sequential governmental actions was evident and immediate, as the amount of ocean dumping of MSS in the ROK became zero in 2012.

Although the ocean dumping of MSS was terminated in 2012, the MOF still allowed temporal dumping of sludge produced from private waste treatment plants, only when ocean dumping was inevitable. However, the MOF continued its efforts toward zero dumping of all kinds of sludge by gradually reducing the total amount of ocean dumping of sludge. As a result, the amount of total ocean dumping was 2016×10^3 m^3 in 2012 when the dumping of MSS was banned, and it was reduced to 900×10^3 m^3 in 2013 when the ocean dumping of food-oriented sewage sludge was banned. In 2014, the MOF implemented a certificate system to permit the dumping of other waste sludge; this has led to a dramatic reduction in the amount of sludge dumping, as it was 352×10^3 m^3 in 2014 and 154×10^3 m^3 in 2015. Finally, in 2016, all kinds of sewage sludge, including wastewater sludge, were banned for ocean dumping in accordance with the establishment of the policy of zero dumping of sewage sludge into the sea in the same year [20].

The environmental conditions at designated dumping sites have improved continuously since 2016. However, they were still contaminated, as the heavy metal concentrations in some of these sites were higher than the threshold effect level (TEL) established by the MOF (Figure 2b). Because the dumping sites were located far away from land and the water depths of the sites ranged from 80 m to 1000 m, the commonly used dredging method for remediating contaminated sediment was not feasible. Since 2016, the MOF has conducted scientific studies on remediating the contaminated sediments. Their results showed that the capping method was effective in achieving this goal [17]. In this method, the contaminated sediments were covered with dredged materials as they were carried to the designated sites from the seashore by vessels.

In Figure 2a,b, the concentration of chromium (Cr) is compared at selected stations in one of the designated dumping sites in the ROK. The concentration was measured in 2019, three years after the capping started in 2016. The stations were selected to compare the Cr concentrations between the capping area and the uncapping area, which was inside the dumping site but did not use capping, and the reference area, which was outside the dumping site and thus Cr concentration in the reference area could be used as a standard. Cr is one of the most serious contaminants and is contained in the sludge generated from leather factories. In the locations shown in Figure 2a, the sludge produced in the wastewater disposal plants near leather factories was dumped and thus the Cr concentration could be used as a measure for contamination.

Notably, the Cr concentrations of two stations (U-1 and U-2) in the uncapping area were higher than TEL, thereby indicating the contamination was serious in these areas. However, the concentration in the capping area (C-1, C-2, and C3) was substantially lower than even in the reference area, which indicated the capping effectively remediated the contaminated sediments in the designated dumping site.

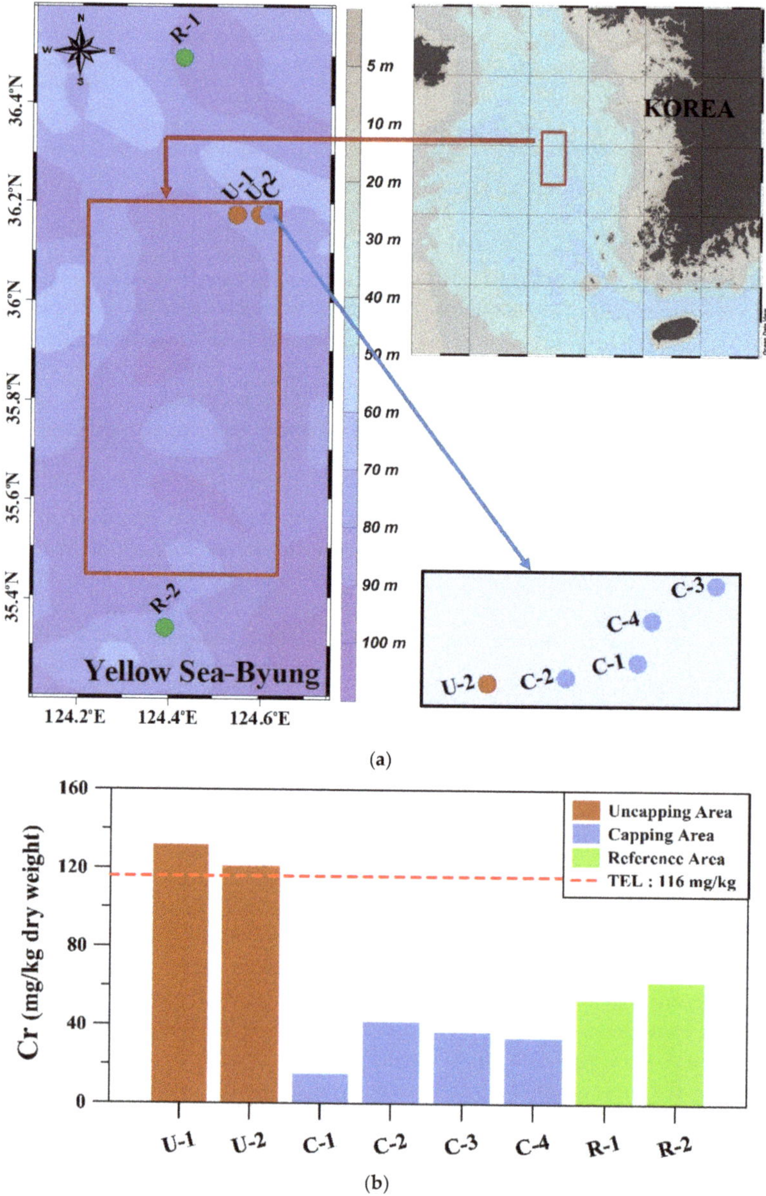

Figure 2. (**a**) Map of the dumping site ("Yellow Sea-Byung") in the Yellow Sea of Korea (red rectangle). C-1, C-2, C3, and C-4 are the locations where the contaminated sediments were covered by dredged materials (capping area), and U-1 and U2 are locations in the uncapping area. R-1 and R-2 are locations selected outside of the dumping site (reference area) and Cr concentration in R-1 and R-2 could be used as a standard; (**b**) Cr concentrations in the surface sediment of selected locations. Blue: concentration in the capping area, Red: Cr concentration in the uncapping area, Green: Cr concentration in the reference area. The red dashed line marks the threshold effect level (TEL) established by the MOF. Cr was chosen for a measure of contamination because as it was contained in the sludge generated by leather factories that was dumped in the Yellow Sea-Byung.

3. Discussion

This review reports the successful outcomes that resulted from the MOF's policies and additionally provides insight for future planning under similar conditions. The initiation of ocean dumping of sewage sludge in the ROK was directly related to the ban of inland treatment/disposal. However, the cheap cost of ocean dumping also contributed to a substantial increase in its usage. Therefore, to reduce the amount of ocean dumping, careful investigation was required to build comprehensive planning that considered all the environmental, economic, and social effects on stakeholders and local communities.

The policies of the MOF that aimed to reduce ocean dumping were successfully applied in the seas of the ROK within a short period, as the Master Plan of Ocean Dumping Management was established in 2006 and the amount of ocean dumping of MSS became zero in 2012. The reason for this rapid success can be found in the application of economic actions such as the polluter-pays principle, which effectively worked for the stakeholders. These actions were strictly applied with fortified surveillance systems such as AIS. In addition, by setting the annual reduction goals of permitted sewage sludge, the amount of ocean dumping could be gradually reduced until it was finally terminated, which provided time for the dumping companies to prepare alternatives. The most successful outcomes in similar scenarios may be potentially achieved from close cooperation between government agencies. The MOF reduced the amount of ocean dumping by increasing the amount for recycling on land through an agreement based on research data and professional advice; additionally, this agreement created socio-economic benefits by encouraging the recycling industry. In addition, the MOF's efforts to minimize conflicts between stakeholders and local communities were successful, as consensus and mutual understanding were reached through continuous education.

This paper focuses on the importance of follow-up monitoring and inspection of the dumping sites even after implementation of the policy and termination of the dumping activities. This is because the ultimate goal of these policies and actions of the government is to recover and maintain the environmental health of the nation. Therefore, studies focusing on determining the best ways to ensure effective policy implementation must be continuously performed to establish future plans or to amend existing policies, and their efficiency should be evaluated based on careful inspection. Regardless of the successful banning of sewage sludge, fish wastes, organic matter of natural origin, and dredged materials are still allowed for ocean dumping in the ROK, and these substances require close monitoring to examine their impacts on the marine environment. In addition, determinations of the environmental impacts of capping methods or determining alternative methods to recover contaminated areas are necessary for investigation of the future safety of the marine environment.

4. Conclusions

Ocean dumping of sewage sludge commenced in 1993 and rapidly increased until 2006 in the ROK as a result of the prohibition on landfills for sewage sludge, which increased negative impacts on the marine environment and raised societal concerns. To properly cope with these impacts, the MOF implemented the Master Plan of Ocean Dumping Management in which ocean dumping of sewage sludge was phased out from 2006 onwards. The MOF's plans and corresponding actions were effective, as they led to termination of ocean dumping of MSS in 2012 and further reduction of the amount of other sludges in the following years until all types of sewage sludge were banned from ocean dumping in 2016. The efforts of the MOF to reduce the amount of ocean dumping have caused an extra effect in that they increased energy efficiency. For example, in 2005, one year before the MOF's plan started, the rate of ocean dumping out of the total mass of sewage sludge in the ROK was 78%, which covered 3/4 of the total amount. In the same year, the rate of recycling of total sewage sludge was only about 5%. When ocean dumping was terminated in 2012, however, the rate of recycling increased to 43%. In addition, the use of energy from incineration and from the collected gas from reclamation significantly increased as well, which contributed to the successful application of the policy of the MOF.

Even though dumping activity was terminated, the MOF has continuously conducted monitoring and research at the dumping sites to understand the contamination process and developed methods to recover the environmental conditions at the sites. Although direct dredging at the dumping sites was not feasible because of the distance from the shore and the water depth, the capping method that covered dredged materials at the top of the area was effective in reducing the concentrations of harmful substances at dumping sites. The success of these policies in the ROK can be attributed to the rapid and strict application of policies with fortified surveillance, in addition to close intergovernmental cooperation and an achieved consensus between stakeholders and communities. However, consistent monitoring is required at the contaminated areas to develop future plans to recover the local marine environment.

Author Contributions: Conceptualization, C.S.C., K.-Y.C., C.-J.K. and Y.S.C.; methodology, C.S.C., K.-Y.C. and C.-J.K.; investigation, C.S.C. and Y.S.C.; resources, C.S.C., K.-Y.C., J.-M.J. and C.-J.K.; data curation, C.S.C., K.-Y.C., J.-M.J. and C.-J.K.; writing—original draft preparation, C.S.C.; writing—review and editing, Y.S.C.; visualization, C.S.C., J.-M.J.; supervision, Y.S.C.; project administration, C.S.C.; funding acquisition, C.S.C. All authors have read and agreed to the published version of the manuscript.

Funding: This research was funded by the Ministry of Oceans and Fisheries (MOF), grant number BSPG51041-12100-2.

Acknowledgments: The authors thank the officers and research team members who were involved in this study, especially those in the MOF and the Korea Institute of Ocean Science and Technology (KIOST). We would like to thank Editage (www.editage.co.kr) for English language editing.

Conflicts of Interest: The authors declare no conflict of interest. The funders had no role in the design of the study; in the collection, analyses, or interpretation of data; in the writing of the manuscript, or in the decision to publish the results.

References

1. Song, G.H.; Choi, K.Y.; Kim, C.J.; Kim, Y.I.; Chung, C.S. Assessment of the governance system for the management of the East Sea-Jung dumping site, Korea through analysis of heavy metal concentrations in bottom sediments. *Ocean Sci. J.* **2015**, *50*, 721–740. [CrossRef]

2. Neto, J.A.B.; de Carvalho, D.G.; Medeiros, K.; Drabinski, T.L.; de Melo, G.V.; Silva, R.C.O.; Diogo, S.; Leandro, B.; Gilberto, D.; dos Santos Filho, J.R.; et al. The impact of sediment dumping sites on the concentrations of microplastic in the inner continental shelf of Rio de Janeiro/Brazil. *Mar. Pollut. Bull.* **2019**, *149*, 110558. [CrossRef] [PubMed]

3. Zhou, Y.; Lei, J.; Zhang, Y.; Zhu, J.; Lu, Y.; Wu, X.; Fang, H. Determining Discharge Characteristics and Limits of Heavy Metals and Metalloids for Wastewater Treatment Plants (WWTPs) in China Based on Statistical Methods. *Water* **2018**, *10*, 1248. [CrossRef]

4. Alemu, T.; Mekonnen, A.; Leta, S. Integrated tannery wastewater treatment for effluent reuse for irrigation: Encouraging water efficiency and sustainable development in developing countries. *J. Water Process Eng.* **2019**, *30*, 100514. [CrossRef]

5. Arguello-Pérez, M.Á.; Mendoza-Pérez, J.A.; Tintos-Gómez, A.; Ramírez-Ayala, E.; Godínez-Domínguez, E.; Silva-Bátiz, F.A. Ecotoxicological Analysis of Emerging Contaminants from Wastewater Discharges in the Coastal Zone of Cihuatlán (Jalisco, Mexico). *Water* **2019**, *11*, 1386. [CrossRef]

6. Yadav, M.K.; Short, M.D.; Gerber, C.; Awad, J.; van den Akker, B.; Saint, C.P. Removal of emerging drugs of addiction by wastewater treatment and water recycling processes and impacts on effluent-associated environmental risk. *Sci. Total Environ.* **2019**, *680*. [CrossRef] [PubMed]

7. IMO. Waste Assessment Guidelines under the London Convention and Protocol, Edition (IA531E). 2014. Available online: http://www.imo.org/en/OurWork/Environment/LCLP/Publications/wag/Pages/default.aspx (accessed on 7 November 2014).

8. Hong, G.H.; Lee, Y.J. Transitional measures to combine two global ocean dumping treaties into a single treaty. *Mar. Policy* **2015**, *55*, 47–56. Available online: https://www.sciencedirect.com/science/article/pii/S0308597X15000184 (accessed on 16 January 2015). [CrossRef]

9. Hong, G.H.; Chung, C.S. *Barriers to Compliance with the Termination of the Ocean Dumping of Sewage Sludge in the Republic of Korea*; Youngjin P&P: Seoul, Korea, 2016; p. 93. ISBN 978-89-444-9042-2.

10. Swanson, R.L.; Bortman, M.L.; O'Conner, T.P.; Stanford, H.M. Science, policy and the management of sewage materials. The New York City experience. *Mar. Pollut. Bull.* **2004**, *49*, 679–687. [CrossRef] [PubMed]

11. Ofira, D.D. The New York Bright years later: Use impairments and policy challenges. *Mar. Pollut. Bull.* **2015**, *90*, 281–298. [CrossRef] [PubMed]

12. Lee, B.G.; Kim, S.W.; Kim, Y.H.; Hyun, C.G.; Lee, H.S.; Kim, K.J. The condition and management measure of marine disposal of wastes. In Proceedings of the KOSOMES Biannual Meeting, 2006; pp. 109–115. Available online: http://www.koreascience.or.kr/article/CFKO200601436758450 (accessed on 16 March 2006).

13. Lee, K.Y.; Chung, C.S.; Kim, Y.I.; Lee, H.K.; Hong, G.H. Contemporary organic contamination levels in digested sewage sludge from treatment plants in Korea: (2) Non-alkylated polycyclic aromatic hydrocarbons. *J. Environ. Sci.* **2005**, *14*, 413–425.

14. Kim, P.G.; Park, M.E.; Sung, K.Y. Distribution of heavy metals in marine sediments at the ocean waste disposal site in the Yellow Sea, South Korea. *Geosci. J.* **2009**, *13*, 15–24. Available online: https://link.springer.com/article/10.1007/s12303-009-0002-8 (accessed on 3 February 2009). [CrossRef]

15. Hong, S.; Shin, K.H. Alkylphenols in the core sediment of a waste dumpsite in the East Sea (Sea of Japan), Korea. *Mar. Pollut. Bull.* **2009**, *58*, 1566–1571. Available online: https://www.sciencedirect.com/science/article/pii/S0025326X09003129 (accessed on 1 July 2009). [CrossRef] [PubMed]

16. Chung, C.S.; Song, K.H.; Choi, K.Y.; Kim, Y.I.; Kim, H.E.; Jung, J.M.; Kim, C.J. Variations in the concentrations of heavy metals through enforcement of a rest-year system and dredged sediment capping at the Yellow Sea-Byung dumping site, Korea. *Mar. Pollut. Bull.* **2017**, *124*, 512–520. Available online: https://www.sciencedirect.com/science/article/pii/S0025326X17306148 (accessed on 14 July 2017). [CrossRef] [PubMed]

17. Jung, J.M.; Choi, K.Y.; Chung, C.S.; Kim, C.J.; Kim, S.H. Fractionation and risk assessment of metals in sediments of an ocean dumping site. *Mar. Pollut. Bull.* **2019**, *141*, 227–235. Available online: https://www.sciencedirect.com/science/article/pii/S0025326X19301407 (accessed on 16 July 2019).

18. Botton, M.L. Effects of sewage sludge on the benthic invertebrate community of the inshore New York Bight. *Estuar. Coast. Mar. Sci.* **1979**, *8*, 169–180. [CrossRef]

19. MOF (Ministry of Ocean and Fisheries). *Development of the Best Practical Technology and Management Options for the National Disposal of Wastes at Sea*; Final report (16th); Ministry of Ocean and Fisheries: Sejong, Korea, 2019.

20. In Public Sewage Statistics, Ministry of Environment. 2013. Available online: www.me.go.kr/home/web/policy_data/read.do?menuId=10264&seq=6564 (accessed on 1 December 2014).

MDPI
St. Alban-Anlage 66
4052 Basel
Switzerland
Tel. +41 61 683 77 34
Fax +41 61 302 89 18
www.mdpi.com

Sustainability Editorial Office
E-mail: sustainability@mdpi.com
www.mdpi.com/journal/sustainability